CABI CONCISE

LETTERS IN ANIMAL WELFARE AND ETHICS

Animal welfare science is relatively young, even compared with other branches of animal science – nutrition, reproduction, animal genetics, etc. Its roots started in the 1980s, stimulated by the intensification of farm animal production during the post-war period. Pioneers of the science emerged, especially in the UK where the intensification had originated, as a result of the need to restore the post-war economy and improve food security. Those of us who joined the march of animal welfare science more recently often fail to realize how hard it was for those early innovators. Derided for pursuing a pseudoscience, ridiculed for suggesting that animals feel pain and that their behaviour is somehow linked to their welfare, these early discoverers charted a pathway for the discipline, which would see it become one of the most popular of the animal sciences.

At the same time as animal welfare science was growing, there was a groundswell of public sentiment supporting more ethical treatment of animals. Whilst animal welfare scientists pondered the meaning of the science, philosophers debated the rights and wrongs of animal ethics and social scientists coined the term 'animal studies' to cement their interest in the field.

This book series, *Letters in Animal Welfare and Ethics*, is about the development of animal welfare science, written largely by those pioneers of the discipline. What were their early thoughts, what unforeseen problems were encountered and how did they navigate the hostile waters of scientific discovery to produce the thriving science that we see today?

Authors have taken their early writings and placed them in the context of the developing discipline, adding their thoughts about more recent discoveries and summarizing the state of knowledge of their particular sector of the animal welfare science discipline.

As areas of academic study, industry concern and public interest, animal welfare and ethics have grown exponentially in recent years. The output of scientific literature dealing with these topics has consequently expanded rapidly.

Highlighting key research and expert voices, this series considers issues as diverse as animal transport within the food industry and the use and misuse of animals in human entertainment. It covers farmed, working, companion, laboratory, sport and wildlife animals, as well as more general areas of discussion such as sentience, personality, welfare assessment and the progression of animal welfare science and animal ethics as academic subjects.

Clive Phillips
clive.phillips58@outlook.com

CABI is a trading name of CAB International

CABI
Nosworthy Way
Wallingford
Oxfordshire OX10 8DE
UK

CABI
200 Portland Street
Boston
MA 02114
USA

Tel: +44 (0)1491 832111
E-mail: info@cabi.org
Website: www.cabi.org

T: +1 (617)682-9015
E-mail: cabi-nao@cabi.org

A catalogue record for this book is available from the British Library, London, UK.

ISBN-13: 9781836992295 (hardback)
9781836992301 (paperback)
9781836992318 (ePDF)
9781836992325 (ePub)

DOI: 10.1079/9781836992325.0000

Commissioning Editor: Alex Lainsbury
Editorial Assistant: Theresa Regueira
Production Editor: Rosie Hayden

Typeset by Exeter Premedia Services Pvt Ltd, Chennai, India
Printed in the USA

Letters on Animal Welfare from the Road to Eden

John Webster

Contents

Preface

This little book arises from an invitation from Clive Phillips to pull out and polish up a selection of my writing over a period of nearly 50 years, during which my scientific career and personal concerns have been directed mostly, though not entirely, to matters of animal welfare.

I was appointed Professor of Animal Husbandry at the University of Bristol Veterinary School in 1977. I had graduated as a veterinary surgeon from the University of Cambridge in 1963, with a Part 2 (Honours) in physiology. My aim then, as always, had been to work with cows, possibly in tropical agriculture and I was advised that I needed at least a PhD to get a worthwhile job. So, after a brief period in large-animal practice, I applied to the Hannah Dairy Research Institute in Ayr, Scotland, for a PhD studentship to study problems of heat stress in cattle. This position had already been filled, however, I accepted an alternative offer from Kenneth Blaxter to do a PhD on cold stress in cattle and, 3 years later, I was appointed assistant professor of physiology in oil-rich Edmonton, Alberta, the coldest big city in Canada, where I was given brand-new lab facilities and access to a lot of cold cows.

This position led to 4 years of high productivity and unalloyed fun, not least the teaching of cowboys whose many skills with cattle put me to shame. However, my Scottish wife, along with our two then very young children, understandably suffered from cabin fever during the long winters, so when I received an invitation from Blaxter, then Head of the Rowett Research Institute in Aberdeen, to lead his energy metabolism department, we decided to return to Scotland. When I was asked at interview whether I was prepared to accept a significant drop in salary, I assured them that 'I was seeking honour without profit in my own country'.

During the Rowett years, my work was primarily focussed on the study of the digestion and metabolism of dietary sources of energy and protein in ruminants, and some of this work involved invasive procedures on animals conducted under licence from the Home Office. At this time, I had no professional involvement in animal welfare and no requests to take up the challenge. My main task was to help develop a new approach to the nutrition of ruminants based on the linked concepts of metabolizable energy and metabolizable protein. This period in full-time research went well but I missed the University life and was looking for an opportunity to get back to teaching when I could enjoy the enthusiasm of young people in particular.

When I applied for the Chair in Animal Husbandry at Bristol, I had sensed the zeitgeist – the explosion of emotional concern for farm animal welfare. Moreover, I had experienced the practice of rearing calves for white veal at first hand and had publicly condemned it on all grounds of health and welfare. Thus my first priority when I came to Bristol was to establish the facilities and a research team to address the most urgent welfare problems arising from intensive production systems for farm animals as addressed by Ruth Harrison (1994) in her seminal book *Animal Machines* on the basis of hard evidence and good science. My aim was to turn down the heat and turn up the light on animal welfare.

The material offered in the following chapters has been selected to reflect the development of my understanding in practical matters of farm animal welfare and fundamental matters relating to the nature of sentience and sentient minds in all animals, including humans. After an enforced retirement from University life, I embarked upon a more ambitious quest – to address the scientific and ethical issues attached to our duty to sustain and protect the living environment. In short, this means extending the principles of good animal husbandry to the husbandry of the planet.

All that follows have been selected from my own previous publications, abridged and polished up a little, to reveal how my thoughts and experience have evolved. Some of my earlier observations have been overtaken by events. However, most of these letters derive from material written after I retired from full-time paid employment – since when I have had more time to think. Some letters (4, 5 and 6) have been taken from lengthy reviews in scientific journals that were supported by comprehensive references to the scientific literature. I present them here in abridged form and support them using a limited selection of only those references that I believe to be essential to convey the message. Readers who seek the bigger picture can refer to the original communications. My intent in these letters is not to overwhelm you with information but to offer 'a good read'.

References

Harrison, R. (1964) *Animal Machines: The new factory farming industry.* Vincent Stuart, London, UK.

Acknowledgements

In my preface, I wrote that my motivation to the academic life was strongly influenced by the desire to teach and enjoy the enthusiasm of young people with lively, enquiring minds. I shall be eternally grateful to all my colleagues and students who ensured that university life was always good fun. It is probably invidious to select just a few names. However, I must record my gratitude and special affection for those most closely associated with these letters: Christine Nicol, Christopher Wathes, Becky Whay, David Welchman, David Main, Nick Bell, Steve Kestin and John Tarlton.

In the Beginning: *Animal Machines* and Ruth Harrison

A Cool Eye

I must begin with an explanation for my title. In my first two animal welfare books, *A Cool Eye towards Eden* (Webster, 1994) and *Limping towards Eden* (Webster, 2005), I gave the word 'Cool' its original meaning. I did this to convey reason over emotion and provide recognition that the concept of Eden as a perfect state where 'the lion shall lay down with the lamb' was unreachable (and wouldn't work anyway) but that it was a good spot on which to focus the eye and a worthy direction in which to travel, albeit in limping fashion. Throughout my career as a veterinary scientist and teacher I have sought to adopt a 'more light, less heat' policy of rational compassion, to the understanding and treatment of animals in our care (more accurately expressed as under our control). Most of this has been taken from the shelves of my earlier works, dusted down and polished for reinspection.

I begin with the opening paragraph of *A Cool Eye towards Eden*.

> Man has dominion over the animals whether we like it or not. Wherever we share space on the planet, which includes all but the most inaccessible regions of the land, sea and air, it is we, not they who determine where and how they will live, we make a pet of the hamster but poison the rat. These are very human decisions, but they have much in common with the decisions taken by other animals, since they reflect our will to survive, preserve our genetic inheritance and enrich the quality of our lives. We need good food and battery eggs are good and cheap.[1]
> We need good hygiene and fear that rats carry germs. Pets enrich our lives and those of our children We admire the tiger not only for its fearful symmetry but as a symbol of freedom itself so we offer it more freedom than we would think fit for a chicken. It is impossible, however, to avoid the issue that both the tiger and the chicken are living on our terms.
>
> (Webster, 1994)

From the earliest days of the hunter/gatherers, throughout the history of agriculture to the present day, humans have cared for animals in direct proportion to their value to us, i.e. their ability to provide sustenance, create wealth and enrich the quality of our lives. In primitive times, this was sustainable. It did not matter too much to the survival of other species because of the limited extent of our dominion over the shared environment. With domestication, however, we have progressively taken away from animals and into our own hands, decisions that we, not they, consider to be necessary for survival and good welfare. At this stage it is necessary only to point out that the few (about 4%) species that have allowed themselves to be domesticated have tended to ensure their survival as a species (if not their quality of life) is rather better than those that stayed in the wild. The chicken is more likely to survive the next hundred years than the tiger.

Today we have complete dominion – sufficient power to destroy the majority of species of 'higher' animals (birds and mammals) but sufficient comfort and wealth to permit us to behave towards (some of) them with compassion and altruism. However, given such power, it is not sufficient simply to feel compassion. If we want to convert our compassion and reverence for life into practical benefit for the survival and quality of life in other species, we need more than the right thoughts, we need the right action.

Since the 1960s, there has been a flurry of books (including mine) on matters of animal welfare. The first to address the abuses of intensive farming was Animal Machines by Ruth Harrison (1964). More of this splendidly pragmatic and constructive book later. Other books, including Animal Rights by Andrew Linzey (1976) and *Animal Liberation: a new ethics for our treatment of animals* by Peter Singer (1990), were, to my mind, worthy but I did not find their concern particularly helpful. It does not matter to them what we think but what we do. Peter Singer attacks humans for the practice of 'speciesism', namely the exploitation of animals for our benefit but without their consent. However, any argument that involves the concept of animal consent is open to criticism on the grounds that the species under consideration cannot contribute to the discussion. To turn Singer's argument on its head, I would suggest that he is guilty of (benign) speciesism by presuming to speak for animals without first pausing (at considerable length) to enquire by all appropriate means a fuller understanding as to what they think and what they feel.

Ruth Harrison

This section is taken from my address *Ruth Harrison – tribute to an inspirational friend* at a conference in Oxford convened by Marian Dawkins in 2014 to celebrate the 50th anniversary of the publication of *Animal Machines*.

The British nation likes to think that it is kind to animals. In 1822, the UK parliament passed 'Martin's Act' to prevent the cruel and improper treatment of cattle, making it the first parliamentary legislation for animals in the world. The most robust single piece of legislation was the Protection of Animals Act 1911, which made it an offence to 'cause unnecessary suffering by doing or omitting to do any act'. This was a good start although the word 'necessary' left too much room for manoeuvre. Laws are written by politicians and, in a democracy, politics is expected to adapt to reflect the changing perceptions of people in a just society. This is usually interpreted as that which is seen as just by the 'reasonable man'. However, to quote Hart (1961) 'it cannot seriously be disputed that the development of the law...has been profoundly influenced both by conventional morality...and by enlightened moral criticism urged by individuals whose moral horizon has transcended the morality currently accepted'.

So often the enlightened moral critics have not been reasonable men but remarkable women, among such notable examples being Elizabeth Fry, Emmeline Pankhurst, Rachel Carson and Ruth Harrison. In her 1964 book *Animal Machines*, which drew public attention to the abuses of factory farming behind closed doors, Ruth Harrison wrote 'If one person is unkind to one animal it is considered as cruelty but when a lot of people are unkind to a lot of animals, especially in the name of commerce, the cruelty is condoned and, once large sums of money are at stake, will be defended to the last by otherwise intelligent people'. This sentence laid bare the mindlessness or wilful hypocrisy necessary to justify so much of intensive farming. In effect, producers and consumers (i.e. nearly all of us) were content to offer a life worth living only to those animals that we got to know as individuals, like our pets. Ruth pointed out the absurdity of the UK Protection of Birds Act (1954) that required any caged bird to be given enough space to flap its wings but then stated, 'provided this subsection does not apply to poultry' (which accounted for more than 99% of all caged birds). Ruth was my inspiration. She was compassionate, yet unsentimental, and was tireless in her pursuit of a fair deal for farm animals. Prior to 1961, she had enjoyed a productive, artistic and essentially urban life but (so far as I can gather) had not displayed any special sentimental attachment to animals. The starting point for her crusade was apparently a single leaflet that drew her attention to the plight of veal calves, broiler chickens and caged laying hens. She said that she was moved to action not primarily from a love of animals but from a burning sense of injustice to those whom she recognized as sentient creatures that deserved to live rather than simply exist before they were slaughtered for our satisfaction.

When I read *Animal Machines*, I was struck by its power. Ruth did not hesitate to present verbal and visual images of the worst of factory farming to maximize the emotional impact of her argument. The picture of the battered hen in the battery cage has done more to influence public opinion than a thousand diligent scientific studies of the welfare of the laying hen. However, Ruth was always meticulously careful to back up her images with evidence. Most of this was obtained by visiting the factory farms and talking with the producers, many of whom had their written and verbal justifications quoted directly and in detail. There was very little animal welfare science in the book for the simple reason that there was very little welfare science to be had at the time. Animal science had been directed almost exclusively to animal production, and I write that as an animal scientist who began his career the year that *Animal Machines* was published. At the end of the book Ruth does not condemn the industry outright. Instead, having set out the evidence, quite fairly, she invites you to make up your own mind – which is easy.

It is a matter of historical record that the publication of *Animal Machines* was the stimulus for the Brambell Report (1965), which reviewed the welfare of farm animals in intensive systems and served to put farm animal welfare as a top priority on the political agenda. Brambell proposed that all farm animals should have, at least, the freedom to 'stand up, lie down, turn round, groom themselves and stretch their limbs'. These minimal standards came to be known as the Five Freedoms (FFs) and for many years dominated discussion of animal welfare in Europe.

I first met Ruth when I was appointed to the Farm Animal Welfare Advisory Committee (FAWAC), later to become the Farm Animal Welfare Council (FAWC). I had recently moved from a full-time research job at the Rowett Research Institute in Aberdeen to take the Chair of Animal Husbandry at the University of Bristol Veterinary School. While at the Rowett, I had been asked to review a programme of research into the mineral requirements of calves reared for the production of white veal – the aim being to ensure the quality of the meat without serious compromise to productive efficiency. This was my first exposure to factory farming at its worst. I concluded that the solution to the veal calf problem was not simply to ensure the supply of just enough iron to avoid anaemia. Everything was wrong! Lack of fibre in the diet caused chronic digestive disorders and these predisposed the calves to secondary pneumonia. Confinement in individual pens caused chronic discomfort and injury. The animals were denied natural oral (rumination) stimulation, comfort and social behaviour. Lacking experience of the normal sights and sounds of farm activity, they panicked at the slightest alarm.

When I attended my first meeting at the FAWC, I was in a state of outrage at the iniquities of white veal production and attacked the Brambell FFs as being a seriously inadequate description of welfare needs. Invited to come up with something better, I proposed the following, which, after some refinement, became the *Five Freedoms and Provisions* (FAWC, 1993):

- freedom from thirst, hunger and malnutrition by ready access to a diet to maintain full health and vigour
- freedom from thermal and physical discomfort by providing a suitable environment including shelter and a comfortable resting area
- freedom from pain, injury and disease by prevention or rapid diagnosis and treatment
- freedom from fear and distress by providing sufficient space, proper facilities and the company of the animal's own kind
- freedom to express normal behaviour by ensuring conditions that avoid mental suffering.

The FFs are considered at greater length in Letter 4.

Here follows a final observation on the state of animal welfare science in these early days. On arrival at Bristol, my first application to the

Agricultural Research Council (ARC as was) was for a study designed to improve the health and welfare of veal calves through the development of more humane, though comparably efficient, alternative husbandry systems. At the time, the ARC rejected my application on the grounds that 'because insufficient is known about this subject at the present time, we feel this work is premature' (sic). In consequence, my first grant for animal welfare science came from the Farm Animal Care Trust, a charity set up by Ruth herself. In later years, the 'reasonable men' of the research councils caught up with this good lady 'whose moral horizon has transcended the morality currently accepted'.

During her time on the FAWC Ruth was always passionate in the pursuit of justice but equally open to reasoned argument and to new knowledge emerging from developments in science and practice. However, she retained her scepticism in regard to comforting assurances. On more than one occasion, she displayed the true courage of the early physiologists by testing things out on herself. Ruth submitted herself to procedures for carbon dioxide stunning and electro-immobilization, both promoted as humane. The first she pronounced terrifying, the second excruciating. We believed her and acted accordingly.

The relentless vigour with which Ruth pursued her campaign was softened, and thereby made more effective, by her charm (she had a splendidly earthy laugh) and her willingness to listen to counter arguments. However, rarely would she accept these counter arguments at face value, whether from the industry or from scientists. Confident in the fact that she was usually as well informed as her interlocutor, she would pick out specific errors and shortcomings, and insist on review and remedy. When we were drafting reviews for the FAWC, getting towards the end of the day and looking forward to the first drink or going home, she would draw our attention to areas where she was not satisfied and where she thought we could and should do better. This could be exasperating, but in my opinion, she was nearly always right.

The Brambell Report (1965) put farm animal welfare as a top priority for the political agenda. More than that, it was the torch that lit the flame of the farm animal welfare movement, which has brought about real improvements in the welfare of farm animals. This has come in part through legislation for improved minimum standards for animals in the intensive systems that caused the most concern – confined veal calves sows and laying hens. However the major force for change has been, and will continue to be, through increased public awareness of the problems going on behind closed doors, demand for improvements and action to reward more humane husbandry through the purchase of quality assured and quality controlled high welfare food from animals. We still have a long way to go but the accelerating pace of improvement has exceeded my expectations. It is a great shame that Ruth died just before the pace of improvement really started to quicken. However, her legacy is immortal.

Note

¹ These words were written in 1994 when battery cages for laying hens were legal in UK

References

Brambell, F.W.R. (1965) *Report of the Technical Committee of Enquiry into the Welfare of Animals Kept Under Intensive Husbandry Systems.* Cmnd.2836, HMSO, London.
Farm Animal Welfare Council (FAWC) (1993) *Second Report on Priorities for Farm Animal Welfare.* DEFRA, London.
Harrison, R. (1964) *Animal Machines: The New Factory Farming Industry.* Vincent Stuart, London.
Hart, H.L.A. (1961) *The Concept of Law.* Clarendon Press, Oxford.
Linzey, A. (1976) *Animal Rights.* SCM Press, London.
Singer, P. (1990) *Animal Liberation: A New Ethics for our Treatment of Animals.* Avon, New York.
Webster, J. (1994) *Animal Welfare: A Cool Eye towards Eden.* Blackwell Science, Oxford.
Webster, J. (2025) *Animal Welfare: Limping towards Eden.* Blackwell Science, Oxford.

Sentience and Sentient Minds

2

Two acts encapsulate our moral and legal responsibilities to animals in our care. The UK Protection of Animals Act (1911) states that it is 'an offence to cause suffering by doing, or not doing, any act'. This is clear but limited. It does not, for example, command us to give regard to their quality of life. The Treaty of Amsterdam (1997) states 'since animal are sentient beings, members shall provide full regard to their welfare requirements'. This expression conveys a broader sense of empathy and compassion, but it lacks precision. The word sentience lacks specificity so has become, in practice one that can mean, in the words of Humpty Dumpty, 'just what I choose it to mean, neither more nor less'. It does not address questions that could be posed by an enquiring eight-year-old (or a lawyer) such as: 'Are all animals sentient?' 'Are some more sentient than others?' 'If so, where do we draw the line?'

From here I shall explore biological and ethical principles that lay the foundations for our understanding of the complex nature of animal sentience and sentient minds. The bedrock of my ethical approach is laid by the immortal words of Albert Schweitzer, 'The great fault of all ethics hitherto has been that they believed themselves to have to deal only with the relations of man to man. In reality, the question is 'what is his attitude to the world and all that comes within his reach? Ethics is nothing other than Reverence for Life' (Brabazon, 1995). The biological principles are grounded in the central truths of Charles Darwin, incontestable because they were built on hindsight deriving from lengthy and meticulous observation. The central tenet of Darwinism is that 'It is not the most intellectual of the species that survives; it is not the strongest that survives but the species that is able best to adapt to the changing environment in which it finds itself' (Bowler, 1996).

Successful sentient species develop the skills that matter most – those that best enable them to ensure their own fitness and that of the environment on which they depend. It also means that they neglect others. All species, including ours, are ignorant. We are just ignorant about different things. It follows that no successful species can claim to be better than any other and there is nothing in Darwinism to justify the anthropocentric belief that evolution has involved a progressive advance in cognitive and emotional development from primitive creatures via the 'higher' mammals to humans at the top of the pyramid. For example, crows are more advanced toolmakers than chimpanzees, the albatross can travel thousands of miles across the featureless southern oceans and return to the same nest yet may fail to recognize a chick that has fallen out and the capacity of dogs to comprehend human speech may only compare to that of a three-year-old but their capacity to acquire and interpret information by scent exceeds our imagination. I should not need to add that in the context of sustained fitness within a sustainable environment, the human race, at the moment, can hardly be defined as a success.

Another meaningless anthropocentric question is 'What are sentient animals for?' Species evolve to promote their own fitness. They were not put on earth by God to serve our needs. It is irrelevant to their needs whether we categorize them as wild (with subsets such as game and vermin) or domestic (with subsets, pet (dog), farm (pig) and sport (horse)). A tiny minority of species (<4%) have accepted domestication, but there is no such thing as a domestic animal. If we are to understand them, we have to imagine the world as perceived through their senses, not ours.

The Five Skandhas of Sentience

While all animals may be sentient, some are clearly more sentient than others. I believe that the most satisfactory *scientific* classification of the varied nature of sentience within the living world is contained within the teachings of the Buddha. He recognizes five categories, or 'skandhas' of sentience, defined (in English) as matter, sensation, perception, mental formulation and consciousness. These are illustrated in Fig. 2.1 as five sections of a cone signifying increasing complexity from the simplest, broadest criterion, defined as *Matter*, to the most select and, complex expression, namely *Consciousness*.

According to the Buddha, all living forms, animals and plants, meet the simplest criterion of *Matter*. I have trouble with this definition since all things that have mass, animate or inanimate, have matter. With due respect to the Buddha, I would prefer to define this most comprehensive description of living matter simply as *Life*. This makes scientific sense since all lives from the simplest amoeba to humankind are dynamic systems, all respond to environmental stimuli and all exploit resources to promote their own fitness.

Sensation describes the capacity of living creatures to interpret stimuli from the external and internal environments (e.g. pain and hunger) as *feelings* that may be sensed as more or less aversive or pleasant. The intensity of these feelings serves as motivation to appropriate action. In the absence of any convincing evidence to the contrary, we may assume this degree of sentience is restricted to animals. We do not yet know how far it extends within the animal kingdom and this is a matter of concern. The Animals (Scientific Procedures) Act 1986 that regulates and restricts procedures likely to cause 'pain, distress,

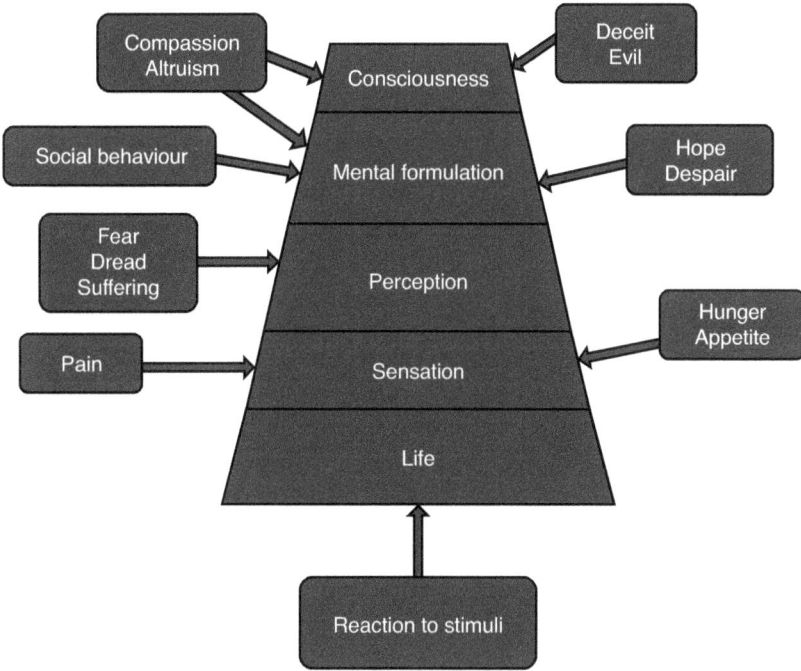

Fig. 2.1. The five skandhas of sentience, with examples of sensations, feelings and emotions linked to each level. (From Webster, 2022, with permission of John Wiley & Sons.)

suffering or lasting harm currently applies to all vertebrates and some invertebrates (including cephalopods like the octopus). This list will grow with time. At this degree of sentience, the recognition of stimuli such as pain, heat, cold, hunger, alarm may involve no more than reflex responses designed to deal with immediate challenge. Nevertheless, in the Scientific Procedures Act it is deemed sufficient to qualify for protection from any human actions likely to cause pain, suffering or lasting harm.

The three inner, more profound, rings of sentience *are perception, mental formulation and consciousness. I shall define these as properties of a sentient mind.*

Understanding the Sentient Mind

The mind, as such, does not actually exist. It is a complex abstraction that we can never fully grasp. The brain exists, which makes it easier for scientists to look into it using expensive bits of kit. We can observe the brain and its operation in ever increasing detail and record the electrical and chemical transmission of information. We can link specific sites and operations in the brain with specific thoughts and actions and provoke these actions with externally or internally applied stimuli. We know a lot about the biochemistry of emotion and the interactions between emotion and cognition – the ways in which how we feel can affect the way we think. All this is fascinating stuff and vital to our understanding of the diagnosis and treatment of mental disorders but, so far, it falls a long way short of creating a coherent explanation of what we imagine to be the mind.

The question, 'What is the mind?', has engaged, employed and escaped the grasp of philosophers and scientists for more than 2000 years. Most of us adopt a more relaxed approach encapsulated by the exquisite phrase 'be philosophical, try not to think about it'. We take it for granted that mind is a useful word to describe how we each interpret the world the way we see it. We might call this our mental state but this is, once again, a circular argument. The majority position today, shared by philosophers and scientists is for *physicalism*, the assertion that mind is not an abstraction but instead that everything in the mind is physical. Reductive physicalists assert that all mental states will eventually be explained in terms of measurable neurophysiological processes. This approach has obvious appeal to computer scientists developing better brains (and minds?) through artificial intelligence (AI). I am more sympathetic to the non-reductive physicalists who argue that while all of the mind may be in the brain somewhere, it will never be possible to provide a complete (or even useful) description of the mind entirely in terms of neurobiology. We shall always need both the psychological and the common sense, every day, approaches to our understanding of the mind if we are to make any practical sense of it at all.

My wish is to understand the sentient mind, especially in non-human animals. I illustrate my approach with a simplistic yet neurobiologically acceptable model of how a sentient animal interprets stimuli and sensations and what motivates it to respond in Fig. 2.2).

The control centre of the animal (all of the mind in nearly all sentient species, may be in the brain somewhere) is constantly being fed with information from the external and internal environment (outside and inside the body). Much information, including our perception of how we stand and move in space, is processed at a subconscious level. In Fig. 2.2, these actions are controlled by stimuli that pass directly from receptors in the central nervous system (brain and spinal cord) via the motor nerves to the muscles that control posture and movement. Having learned to walk, we are able to control our limbs without recourse to thought or emotion. Many animals, including raptor birds and top cricketers, have an exquisite ability to process the movement of an object in flight and programme their own movements so as to catch it. This is an amazing property of the neuromuscular system, but the only 'conscious' element was the decision to go for the bird/ball in the first place. The decision to act (or not) in response to a stimulus must involve some degree of interpretation. The brain of all sentient animals is equipped at birth with a foundation programme for survival constructed from the specific gene-coded information acquired by its ancestors through generations of adaptation to the challenges of the environment in which they evolved. I refer to this property of mind

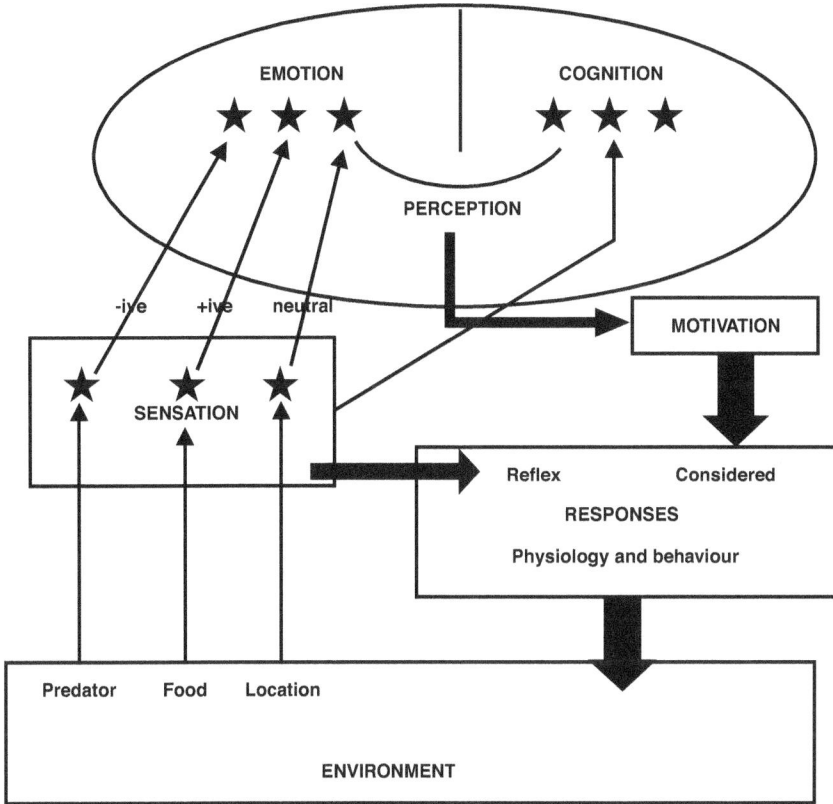

Fig. 2.2. The sentient mind. (From Webster, 2022, with permission of John Wiley & Sons.)

as their *mental birthright*. For animals that demonstrate only the property of sensation, this may define the limits of their sentience. For animals that demonstrate the three inner circles of deep sentience, this foundation programme is expanded and enriched over time as a result of learning and experience. In animals with the power of perception, an incoming stimulus does not enter an empty room but becomes just another guest at the party, one that may or may not make an impact.

In Fig. 2.2, signals from external sensors such as the skin, the eye and the ear, and those from internal sensors for example, biochemical parameters such as blood sugar, are recognized by the primary receptors as sensations. Much of the information from external sensors such as the skin and the eye will invoke a subconscious, reflex response such as immediate withdrawal from a source of pain or threat. However, signals that are interpreted by the primary receptors as sensations will also pass to a second set of neural receptors linked to emotion (how it feels) and cognition (how it thinks). Negative stimuli like pain and hunger will be interpreted in the emotion centre as bad feelings and motivate the animal to conscious action to remove the source of pain or seek food. Other information, such as location or time of day, may be interpreted simply as information with no emotional component and pass directly to the cognition centre. The response of the brain to incoming sensations and information may or may not involve a cognitive element but the motivation to conscious (as distinct from reflex) behaviour will nearly always have an emotional component. Indeed, I would argue that this is the essence of sentience.

To illustrate this concept, consider the primitive but clearly conscious sensation of hunger. The appetite control centre, which can be accurately located in the brain, will monitor internal stimuli such as low blood glucose or incoming sensory information from hunger pangs arising in the gut. It will also respond to natural external stimuli, such as the arrival of food, or a conditioning stimulus, such as the bell that Pavlov rang to condition dogs to anticipate the arrival of food. This direct information, like all incoming stimuli, will be interpreted in the context of existing food files in the brain carrying information from past experience as to the palatability, or possible poisonous nature of the food, or, if no food is currently on offer, the likely time to the next meal. If the animal is hungry and no food is available, it will experience a negative emotion that may range from slight to severe depending on how hungry it feels. When food arrives, it will feel good to a degree depending on the intensity of its former negative state (how hungry it felt) and the appeal of the food on offer. If the animal is already full, it may treat the arrival of another meal with indifference and the information will be interpreted as neutral. My cats are masters of this studied indifference.

The notion that complex incoming information can be categorised on an emotional basis simply as positive, negative and neutral may seem overly naïve but it is supported by classic experiments in neurobiology (Kendrick, 1991). Direct recordings from nerves in the brains of sheep reveal that when they are presented with food, or even pictures that indicate food (e.g. sacks of grain, bales of hay) these trigger signals in neurones that convey a positive message (a good feeling). Other images, such as pictures of dogs or humans, trigger a negative message (danger). However, when the sheep were shown an image of a familiar human carrying a bucket of food, these two categories of information (food and potential predator) were processed at the emotional level and passed on as a simple, unconfused positive message, 'I feel good'.

Animals may need only the first two circles of sentience to experience hunger as a sensation that increases in severity the longer they go without food and so increases the strength of motivation to seek a meal. Animals that possess the third category of sentience, namely perception, can adjust their actions in the light of their past experience, choosing, for example, high energy foods that they associate with greater satisfaction in the past and avoiding foods that they remember made them feel sick after the meal. The message at this stage is that an animal with the faculty of perception can remember the nature and intensity of sensations that induce positive and negative emotional states and will be motivated to actions designed to make it feel better at the time and avoid circumstances likely to make it feel bad in the future. At this level, sentience can truly be described as 'feelings that matter'.

In Fig. 2.2, I have located the cognition and emotion centres on the left and right sides of the brain (we are looking at it from the front). Although this is overly naïve in an anatomical sense, it is consistent with current understanding of *lateralisation* within the conscious reception and control centres in the brains of vertebrates (Rogers, 1980).

In mammals, the cerebral cortex is divided into two distinct halves, with some linkage via the corpus callosum. Nearly all sensory and motor nerves cross over before entering and after leaving the brain so that the left side receives sensation from and controls muscles in the right side of the body and *vice versa*. So far, straightforward. The really interesting feature of lateralisation is that the left and right sides of the brain interpret incoming information and sensation in different ways. In humans, the left side of the brain appears to be primarily involved in the processing of information in detail, the right side with getting an overall feeling about the big picture. Humans with damage to the left brain have an impaired ability to understand and use language. Malfunctions in the right brain are associated with disorders of emotion such as depression and delusions. All this reinforces the notion that the sentient mind possesses the properties of both emotion and cognition. Sometimes, it operates on the basis of both feelings and thought, but always on at least the basis of feelings. This concept was central to the thinking of the psychiatrist Iain McGilchrist who entitled his investigation of the divided brain '*The Master and his Emissary*', the master being the right, emotional side holding the big picture, the left being the well-trained civil servant with the learning and skills needed to do the right thing (McGilchrist, 2012). I should add that Plato was there first with his allegory of the mind as a charioteer driving two horses, passion and logic.

The relevance of lateralisation to our understanding of the sentient mind in non-human animals is beautifully illustrated by the behaviour of birds, who have no corpus callosum so minimal links between the left and right brain. Vision, sensory input from the eyes, like nearly all other sensory information, crosses over *en route* to the brain. Thus, information from the right eye is processed in the left brain and *vice versa*. When chickens are foraging for food, seeking information, they tend to favour their right eye, so transmit specific information to the left side of the brain where it is interpreted in detail. When on the look-out for, or under threat from a predator, they favour the left eye, so stimulate a broad emotional response, fear, fight or flight (Rogers, 1980).

To summarize the argument so far: the sentient mind operates primarily on the basis of feelings, modified to a varying degree by cognition, simply expressed as the ability to think about sensations and emotions. The power of cognition is widespread throughout the animal kingdom. We already know a lot about the cognitive abilities of mammals and birds and are discovering more and more evidence of the capacity of fish and invertebrates like cephalopods to construct complex mental formulations. The point I wish to emphasize at this stage is that, for sentient animals, including humans, most of the time, cognition is acting in the service of emotion.

I make no claim that this is a novel insight. Observations from David Hume, from his 1740 Treatise on Human Nature are of note: 'Reason is, and ought only to be, the slave of the passions and can never pretend to be any other office than to serve and obey them. The causes of these passions are likewise much the same in beasts as in us, making a just allowance for our superior knowledge and understanding' (Hume, 1740).

Later philosophers (except for Darwin) have quarrelled with this bleak, animalistic view of nature. Humans may argue, without clear evidence, that we are better than this because we alone possess the fifth skandha, namely consciousness, which is the facility of being aware that we are aware. Clearly this has enabled some of us to plan and do clever things both good and evil. However, we evolved as sentient creatures and the evidence strongly suggests that we too, most of the time, are motivated primarily by how we feel, whether our considered actions be selfish or altruistic. This is a good reason why I am confident that AI will never reproduce the human brain. For sure, it can make information processing computers that are better informed and quicker thinking than we have currently, leading to amazing developments such as creating better antibiotics, developing gene-specific cures for rare cancers, or playing better chess. However, I do not envisage an artificial brain driven by how it feels. Scientists will, I am sure, develop an artificial brain that *appears* to act according to how it feels. I suggest that this will be no more than a brain that acts according to how they feel it should feel, which is not the same thing.

Animals do not require the deepest circle of sentience, namely consciousness, in order to interpret and remember past experiences and the feelings they arouse. Thus, they cannot be said to live only in the present. Their emotional state will be defined by their expectations of the future in the light of past experience. This has profound implications for our understanding of the impact not only of primitive emotions such as pain and fear but also so-called 'higher' feelings such as hope and despair, comfort and joy.

Properties of the Sentient Mind

Table 2.1 outlines some of the main emotional and cognitive expressions of sentient minds.

The power of perception gives animals the ability to adapt to sensations such as pain, fear and hunger.

Fig. 2.2 illustrates how the sensation of fear is perceived not just as an emotion and a stimulus to action but also as a learning experience. An animal that learns that it can take effective action will habituate, i.e. it will develop an increased sense of security. If it learns that it cannot take effective action – either because the stresses are too severe, too complex or too prolonged, or because it is confined (usually by us) in an environment wherein it not possible to take effective action – then it will suffer.

Table 2.1. Emotional and cognitive expressions of sentient minds. (Author's own table.)

	Emotion	Cognition
Perception	Pain and fear	Avoidance
	Hunger and thirst	Food selection
	Comfort	Nest building
	Curiosity and security	Interpret simple social signals
Mental formulation	Anxiety and depression	Understand social signals
	Pleasure, joy, hope, grief	Education and culture
Consciousness	Affiliative behaviour	Aware of self and non-self
	Empathy and compassion	Deceit

The power of mental formulation gives animals the capacity to do more than simply learn by rote. It gives them an understanding of cause and effect, which enables them to develop coping strategies, not only as individuals but through observing and understanding the actions of others. Parents educate their offspring and communities develop a culture. There is good evidence of education and cultural development in a wide range of mammals and birds. Among the mammals, there are proven examples in primates, cetaceans (dolphins and whales), elephants and wolves. Birds that display evidence of cultural development include corvids, herons, songbirds and even the humble domestic fowl. (For further discussion of these and other examples, see Webster (2022) or my website websterwelfare.com.)

The power of consciousness, as defined by the skandhas, equates to a human sense of 'being aware that we are aware'. This is usually equated with a sense of self and non-self and from this, the ability to consider the thoughts and feelings of others. This enables the expression of positive emotions and actions such as empathy and compassion. It also conveys the capacity for emotionally and cognitively complex forms of anti-social expression such as deceit and evil.

Expressions of Perception

Suffering

We humans talk of 'suffering pain'. In science, pain has been described as 'an unpleasant sensory and emotional experience associated with actual tissue damage'. This definition recognizes both the elements of sensation and emotion. Too many physiologists in the past have ducked questions concerning the emotional elements of pain, preferring simply to consider nociception, the sensation of pain. Some have argued that pain, even in humans, is such a subjective experience that it is not open to scientific explanation. To answer them, I recruit Wittgenstein 'just try, in a real sense, to doubt someone else's fear and pain' (Wittgenstein, 1953). I am happy to extend that doubt to other sentient species. In what follows, I shall adopt the simple model illustrated by Fig. 2.2 and describe sensations and emotions associated with pain simply as positive or negative, in full recognition of the fact that these bald terms lack subtlety of meaning. I shall use 'mood' when I refer to emotional responses. Many scientists prefer the word 'affect'. I can think of no good reason why.

The possession of a sentient mind brings with it the capacity to suffer. This feeling may be modified by cognition. For example, the mood of a woman with severe abdominal pain will differ according to whether she knows she has cancer or is giving birth to a baby but the sensation may be the same. We can say with some certainty that potential sources of suffering in non-human animals include pain, fear, hunger and thirst, severe heat and cold, malaise (feeling ill) and exhaustion. We can say with equal certainty that we humans can suffer from anxiety and depression, boredom and frustration, loneliness and loss. We have a duty to explore the extent to which we may share these 'higher' emotions with other sentient species.

We can describe our own experience of pain as a negative experience, a bad thing, both in terms of the immediate sensation and its emotional consequences. We seek to avoid sources of pain, we take drugs to reduce the sensation and, if the pain persists, our mood takes a turn for the worse. In recent years, there have been many excellent, compassionate studies of pain as a sensory and emotional experience with mammals, birds and fish that convince me of their capacity to suffer.

My personal links with this largely involve the work of colleagues at the University of Bristol with broiler chickens (Knowles *et al.*, 2008). I shall briefly summarize this work as an illustration of a more general theme. The word 'broiler' describes chickens that are selectively bred and reared to produce meat as quickly and cheaply as possible. Modern broiler strains can progress from the moment of hatching to the moment of slaughter in 40 days or less. As a result of this, extreme pressure for the increased productivity of an increasing number of birds has developed in these birds a condition the trade has called, without shame, 'leg weakness'. As they approach slaughter weight the birds become increasingly reluctant to move, many go off their legs altogether and die or have to be culled, not least because they can no longer reach the feeders. The pathology that gives rise to leg weakness is complex. It can involve both abnormal bone growth and joint damage, often exacerbated by infection, but it is all precipitated by the fact that these birds are outgrowing their strength. There are serious issues of animal welfare and human ethics arising from the problem of leg weakness in broilers. These have been addressed at length elsewhere and it is fair to say that the industry, largely in response to consumer pressure, has gone some way to putting its house in order. However, this is outside my current brief, which is strictly concerned with the extent to which the pain associated with leg weakness in strains of these birds that have been selected for rapid growth may be a source of pain and suffering.

I will start from the extreme position of the devil's advocate. It would be possible, in the absence of further evidence, to claim that the progressive loss of mobility in fast-growing strains of broiler chickens could simply reflect increasing mechanical interference with joint movement, possibly accompanied by an imbalance between muscle strength and body weight, which is not necessarily accompanied by pain. Others have argued that chickens may experience the sensation of pain unaccompanied by any element of emotional distress and so not actually suffer.

We cannot simply dismiss these arguments as heartless, we must construct a solid, evidence-based case if we are to show that chickens can experience pain and that this pain causes them to suffer. Learned avoidance of a behaviour likely to cause pain, such as jumping down from a perch, provides some evidence for an emotional response. Changes of mood, such as reduced appetite or grooming, or positive actions such as exploration and play, offer further, but not yet totally convincing, evidence that pain has an emotional element in the species under observation (here, the chicken). The most convincing evidence that animals suffer when in pain comes from experiments that involve rats, chickens and several other species of sentient animals self-selecting analgesic (pain-killing) drugs in their food or water. In these trials, it is important to avoid drugs that induce euphoria, such as opiates, since all animals tended to favour and become addicted to these. With non-euphoric analgesics such as the non-steroidal anti-inflammatory drugs (NSAIDS), animals without damage likely to cause pain tended to prefer the food without the drug, probably because the treated food tasted strange. However, in this case the chickens that displayed signs of abnormal locomotion preferentially chose the food containing analgesic and consumed enough of this food to administer what the manufacturers would advise to be the correct analgesic dose. They have perceived how to relieve their pain through self-medication. To my mind, this is the final, convincing link in the chain of evidence that proves that, in these species, pain is more than just a matter of sensation. We may safely conclude that when they are in pain they suffer.

Fear and dread

Fear is an essential, highly functional, primitive sensation that acts as a powerful motivator to behaviour designed, where possible, to avoid threat. It is also an educational experience since the memory of previous threats, the action taken in response to those threats and the consequences thereof ('was it less bad than I feared or worse?') will obviously affect how the animal feels next time around. Causes and consequences of fear are illustrated in Fig. 2.3, which identifies three main threats, novelty threats, innate threats and learned threats.

Neophobia, or the fear of novelty, is an obvious survival mechanism. Success in life depends on achieving the right balance between curiosity (to develop survival skills) and caution (to avoid danger). Among these skills is the ability to learn the distinction between threats that are real or imagined. Most innate fears (e.g. primates' fear of snakes) have survival value. Arachnophobia (humans' fear of spiders) would appear to be a lasting remnant of our primitive

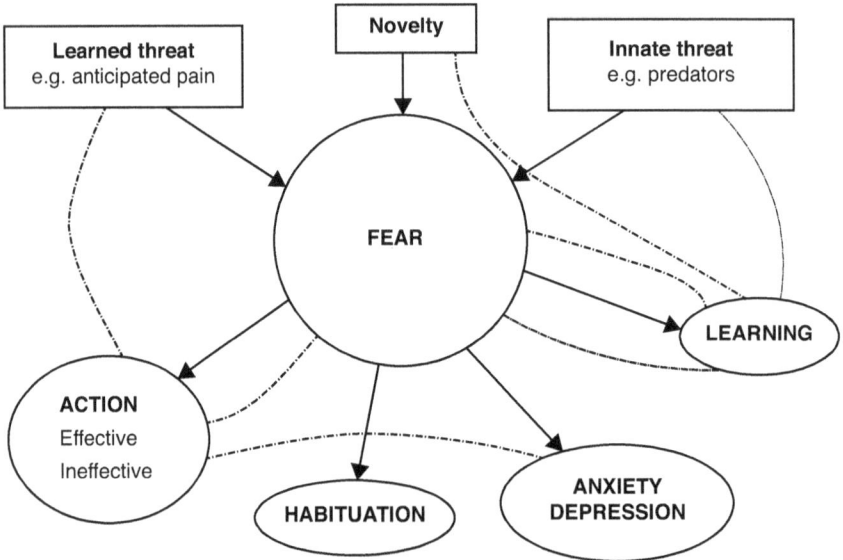

Fig. 2.3. Fear, threats, reactions and consequences. (From Webster, 2022, with permission of John Wiley & Sons.)

past. While these innate threats are relatively hard-wired (i.e. built into the mental birthright) they can be overcome by experience. Fear of a learned threat is self-evidently one that is acquired by experience, e.g. the fear shown by many dogs on visits to the vets. The dog that is returned to boarding kennels prior to the annual family holiday, or even watches the family packing up, may experience a more advanced form of learned fear, i.e. that of desertion by key members of its social group.

Stimulated by fear, the animal takes what action it can. If it learns that its action has been successful, it will know what to do next time. It has discovered that it can cope and will experience an increased sense of security. If it discovers that it cannot resolve the problem, either because its actions were ineffective or because it is in an environment that restricts its ability to act, then it may suffer a chronic, non-adaptive change in its emotional state or mood, within a spectrum ranging from high anxiety to severe depression. In simple terms, acute fear morphs into chronic dread.

Mental Formulation

I repeat, 'the power of mental formulation gives animals the capacity to do more than simply learn by rote'. They acquire an understanding of cause and effect, which enables them to develop coping strategies, not only as individuals but also through observing and understanding the actions of others. Some of the most convincing examples of animal behaviour that would appear to require the capacity for mental formulation are revealed by expressions of social behaviour. These include the:

- establishment of family bonds that extend beyond the immediate, instinctive needs of feeding offspring to the point of weaning or fledging
- establishment of special relationships outside the immediate family, reinforced by affiliative behaviour such as grooming
- formation of alliances within the group for assistance in encounters with outsiders
- unselfish expressions of empathy and compassion for others, such as consolation behaviour to one who may have lost an offspring or come off worse in a fight.

The expression of these behaviours in different species is determined to a great extent by their birthright, itself laid down by the genetic consequences of many generations of adaptation of their ancestors to the environment. Sheep and horses are herd animals, tigers are solitary, wolves and elephants live in extended family groups. However, all these varying patterns of social behaviour reflect the same motivational drives. On a population basis, successful behaviour is that which promotes genetic fitness, which is a hard-wired (mindless) aim. However, the genetic fitness of the population depends upon the behaviour of individuals, who are motivated not by an understanding of Darwin but instead by their immediate individual and family needs. These will always include primitive needs for security, sex and survival but, in species with the capacity for mental formulation, may also include 'higher' emotions such as friendship, empathy and compassion.

Most herbivores, being prey species, live in large herds. At the simplest level, the success of this behaviour does not require a degree of sentience deeper than that of hard-wired sensation. The probability of one sheep being killed by a passing wolf is obviously greater if it is out on its own than if it is within a large flock. However, there is good evidence for more highly-developed flocking behaviour in herd animals. Musk oxen in the Arctic, with calves at foot, defend themselves against attack from a wolf pack by forming a circle, with the aggressively horned adults pointing outwards and the calves safely in the middle. The primary motivation for this behaviour is the hard-wired, instinctive need for survival of the group (not necessarily the individual), but its complexity indicates that this is a strategy that has benefited from social learning.

Predator species tend to be solitary, like the tiger, or operate in family groups, like wolves. Again, the proximate reason for this is obvious. The tiger needs no help to stalk and kill its prey; its greatest threat is from another tiger. Wolves are successful when they seek and destroy in packs. Conspecifics outside the family groups are likely to be seen as competitors and a potential threat to survival and genetic fitness.

The primary motivation to social behaviour in a herding, prey species is the need to achieve security by reducing the risk from predation. However, animals compelled to live in close proximity to one another in large groups also need to establish a *modus vivendi* that ensures a quiet life. Most achieve this through the establishment of a hierarchy, or pecking order, which may involve aggression in the early stages but usually settles down unless the group is further disturbed, e.g. by new arrivals. Within these groups, some animals may establish especially close contacts with chosen individuals, who may or may not be related to them, and reinforce these contacts with bonding behaviour such as social grooming. This does not necessarily imply friendship. It may just be a way of increasing strength in numbers when faced by aggressive neighbours. Much of this behaviour is likely to be instinctive and may not require a greater depth of sentience than simple sensation. However, the capacity for mental formulation have not only the capacity to learn from experience but also to educate their offspring and others in their social group. Habits acquired by experience, rather than encoded within the genetic birthright, are passed on from generation to generation. Such groups develop a culture.

We would, I think, describe friendship and love (with hate and evil as the polar opposites) to be expression of the inner circle of sentience, namely higher (human) consciousness. There is, however, good evidence of friendship in animals based on observations of behaviour that include strong affiliative bonds with special individuals, inside or outside the family, and grief at the death or departure of a comrade. Many animals display signs of distress after loss of contact with preferred individuals. Donkeys are particularly likely to suffer severe distress after the loss of a long-time friend and will not be consoled by contact with others. We have, I believe, sufficient evidence to include friendship within the scope of emotions that we should recognize and respect in at least some of the sentient animals. Whether these things may be interpreted as expressions of higher consciousness or, more simply, the capacity for mental formulation is a matter of opinion (and is probably a distinction that doesn't much matter.) The motivational basis of these emotions is uncertain, even in humans. Many stern behaviourists argue that they can be explained simply in terms of the genetic imperative, classically expressed by J.B.S. Haldane who reportedly said, 'I would be prepared to lay down my life for two of my children or eight of my second cousins'. The grief expressed by a donkey, or an elephant, at the death of a close companion may simply be an expression of personal loss. According to the Haldane argument, true love can only be expressed in terms of true altruism, namely behaviour designed to benefit an individual or group outside one's genetic pool. There is some evidence for true altruism in animal societies. Chimpanzees, bonobos and rooks(!) have been observed to offer comfort to others outwith their immediate family showing signs of distress.

Comfort and joy

There is more to the behaviour of animals with sentient minds than the basic need to obtain nutrition and ensure the survival of the species. The chicken that sleeps at night on a perch, out of the reach of rats and foxes, is motivated by an instinctive need for security. However, the cat or lamb, that basks in the sun, or the dog that lies so close to the fire that it gets too hot and starts to pant are clearly motivated by the desire for a strictly unnecessary experience that we may interpret as hedonistic pleasure. This brings us to the question of play. Here again, the stern behaviourists argue that play, as observed in young carnivores, is a strictly functional pursuit. Cubs or kittens that spring out on each other and engage in mock fights may well be learning life skills but that does not exclude the possibility that they are also having fun. Other forms of play behaviour such as lambs or foals 'pronking' (bouncing on all four feet) or engaging in crazy races would appear to have no function other than the pursuit of pleasure.

I was once asked to appear on a television programme to address the question sent in by a viewer 'why are cows so boring?' I explained that the dairy cow, the most overworked of mothers, has neither the time nor the energy for play. I then dropped a couple of balloons into a pen full of calves and at once initiated a game of nose ball. Play exists for sentient animals as a luxury pursuit open to those who have both the time for play and sufficient food to provide the energy for play. In former centuries, affluent English gentlemen, strongly advised not to go 'into trade' and having, in consequence, a surfeit of spare time themselves, devised an impressive range of sporting activities. This is a classic example of the definition of play as a luxury pursuit in a sentient species.

Hope and despair

We can accept that the capacity of perception means that animals suffering the consequences of primitive emotions such as fear and pain provides convincing evidence that they do not live only in the present, but can we extend this to include things that are normally considered expressions of higher human consciousness, namely hope and despair? Here again, I quote the authoritative voice of Wittgenstein, 'One can imagine an animal angry, frightened, unhappy, happy, startled. But hopeful, why not? Hope is an expression of belief, but belief is not thinking' (Wittgenstein, 1953). If we interpret Wittgenstein's assertion about hope in the context of sentience then hope becomes a positive emotion, a feeling that through my actions or the actions of others, something good may ensue and I shall feel better.

It may be an anthropomorphic error to view the concepts of hope and despair as we understand them to be the products of a conscious mind, aware that it is aware. Human development of the concept of hope may be much more complex than that of other animals and may extend, not always rationally, much further into the future, viz. 'the sure and certain hope of the resurrection to eternal life' as expressed in the Anglican Book of Common Prayer. However, I have no problem in defining hope simply as an emotional feeling that the state of feeling good can be sustained or may, in future, improve. A simple illustration of the concept of hope in animals is provided by the primitive sensation of hunger. Calves that bleat, sows that chew the bars, or wild cats that pace their cages in anticipation of a meal are displaying an emotion linked to the expectation that a meal will arrive at the expected time. When the hoped-for event arrives, their behaviour is consistent with pleasure.

The concept that hope is a learned emotion, a feeling about the future based on past experience, is particularly relevant to domesticated animals since the pattern of their lives is so dependent on the actions of their owners. If a dog or farm animal is fed at increasingly irregular intervals or not at all, this confident expectation will be replaced by anxiety and this negative emotion will be greatest at former mealtimes. Similarly, the dog that is repeatedly disturbed by the non-return of its owner at the expected time may develop prolonged and incurable separation anxiety. The primate confined for years in a barren cage with little or no social contact will develop the signs (or non-signs) of profound apathy. These observations are reinforced by hard evidence from controlled trials with laboratory animals. Prolonged isolation or social challenges leading to defeat can inhibit neural development in mice leading to lasting changes in physical and behavioural development, such as decreased appetite, growth rate and signs of anxiety or apathy indicative of a profound deterioration in mood. The word to describe the emotional consequences of these failures of hope has to be despair.

Higher Consciousness

The power of consciousness, as defined by the skandhas, equates to a human sense of 'being aware that we are aware'. This is usually equated with a sense of self and non-self and, from this, the ability to consider the thoughts and feelings of others. This enables the expression of positive emotions and actions such as empathy and compassion. It also conveys the capacity for deceit. One classic approach to the search for evidence of 'human' consciousness in animals is the *Mirror test* in which primates, for example, are given a mirror and become accustomed to looking in it. A mark is then put on the face of the animal. If the animal then tries to wipe it off, we conclude that it recognizes the face in the mirror as its own. This is a good test of recognition of self, and this property appears to be mainly restricted to some primates. However, it tells us little, if anything, of the capacity of species, or individuals within species, to express emotions such as altruism, compassion, love, or hate. Indeed, I suggest that it is simply a matter of semantics as to whether we put these properties within the category of consciousness or mental formulation. Whether any non-human species possess the two most human expressions of higher consciousness, 'the knowledge of good and evil', I simply do not know but, in the absence of evidence to the contrary, I would guess not.

Summary

The Buddhist five skandhas of sentience provide an elegant, biologically accurate system for classification of sentience within the animal kingdom. We can assume with confidence that humans possess all five skandhas, but what about other animals? How sentient are they and how much should we care? What about the amoeba? Because it reacts to stimuli likely to cause damage, does this mean it experiences sensation? It would be comfortable to avoid the issue altogether, an approach, described, tongue in cheek, by Roger Brown:

> How much of the mentality that we offer one another ought we to allow the monkey, the sparrow, the goldfish and the ant? Hadn't we better reserve something for ourselves alone, perhaps consciousness? Most people are prepared to hold the line against animals. Grant them the claim to make linguistic references and they will be putting in a case for minds and souls. The whole phyletic scale will come trouping into heaven demanding immortality for the tadpole and the hippopotamus. Better be firm now.
>
> (Brown, 1958)

I am not prepared to be so dismissive as Roger Brown, though I do like his turn of phrase. My aim is to lay the foundations for our ethical approach to animals based on the best available evidence as to the nature of their minds. I believe that sentience begins to matter when it includes the second skandha, sensation. This is consistent with the Animals (Scientific Procedures) (1986), which recognizes that the property of sensation alone is sufficient to give protection to named species from harmful procedures conducted in the interests of science, health and safety. For species with *sentient minds* that possess the powers of perception and mental formulation, our responsibility extends beyond the maxim 'do no harm' and should accept our responsibility to respect their quality of life. This applies whether or not we believe they can experience the 'higher' elements of consciousness that most regard as unique to the human species. We have (I hope) moved on from the Cartesian belief (*Cogito ergo sum*) that only humans have souls because only humans are capable of rational thought. We too are sentient beings, with special skills.

One, final cheeky thought is this: it has been suggested that the claim that we are the only rational species is just rationalisation.

Note

[1] This letter has been largely drawn from Chapter 2 'Sentience and the Sentient Mind' in *Animal Welfare: Understanding Sentient Minds and Why it Matters* (Webster 2022). This comes with comprehensive references to original communications.

References

Animals (Scientific Procedures) Act (1986). Available at: https://www.legislation.gov.uk /ukpga/1986/14/contents

Bowler, P.J. (1996) *Charles Darwin: The Man and his Influence.* Cambridge University Press, Cambridge.

Brabazon, J. (1995) *Albert Schweitzer: A Biography.* Syracuse University Press, New York.

Brown, R. (1958) *Words and Things: An Introduction to Language.* Syracuse University Press, New York.

Hume, C. (1740) *Treatise on Human Nature Book II.*

Kendrick, K.M. (1991) How the sheep's brain controls the visual recognition of animals and humans. *Journal of Animal Science* 69, 5008–5016.

Knowles, T., Kestin, S.C., Haslam, S.M., Brown, S.N., Green, L.E. *et al.* (2008) Leg disorders in broiler chickens, prevalence, risk factors and prevention. *PLOS One* e1545I. Available at: https://journals.plos.org/plosone/article?id=10.1371/journal.pone. 0001545

McGilchrist, I. (2012) *The Master and his Emissary. The Divided Brain and the Making of the Western World.* Yale University Press, New Haven.

Protection of Animals Act (1911) *c. 27.* Available at: https://www.legislation.gov.uk/ ukpga/Geo5/1-2/27/contents

Rogers, L. (1980) Lateralisation on the avian brain. *Bird Behaviour* 2, 1–12.

Treaty of Amsterdam amending the Treaty on European Union, the Treaties establishing the European communities and certain related acts (1997). Available at: http://data. europa.eu/eli/treaty/ams/sign

Webster, J. (2022) *Animal Welfare: Understanding Sentient Minds and Why it Matters,* UFAW Animal Welfare Series. John Wiley & Sons, Oxford.

Wittgenstein, L. (1953) *Philosophical Investigations.* Blackwell, Oxford.

The Ethical Matrix: Beneficence, Autonomy and Justice

Loving animals is not enough. If we are to do right by them, we need to understand them. On the other hand, loving animals can be too much. Our respect for the 'rights' of sentient animals has to be placed within the context of our own legitimate needs. This calls for some ethical decisions. For most of history, moral philosophy, or the study of ethics, has concerned itself exclusively with people. This anthropocentric tunnel vision was challenged by Albert Schweitzer, who wrote 'Ethics is nothing other than Reverence for Life. Reverence for Life affords me my fundamental principle of morality, namely, that good consists in maintaining, assisting and enhancing life, and to destroy, to harm or to hinder life is evil' (Brabazon, 1995).

If we are to adapt principles of ethics within human society into humane and effective action with respect to sentient animals, we need some ground rules. As with all sciences, ethics may be divided into the pure and the applied. Pure ethics is top-down. It asks the question 'Which general moral norms for the guidance of moral behaviour should we accept and why?' The aim of this approach is to justify moral norms. This is a worthy pursuit but one that may have little impact on the behaviour of society at large. Applied ethics is bottom-up. This is a more pragmatic business that begins with life as it is. It identifies a specific practical issue, then constructs an analysis of relevant moral issues by a process of induction. Beauchamp and Childress proposed a practical, bottom-up approach to problems in medical ethics in the form of an 'Ethical Matrix' built on three aims of common morality, to promote wellbeing, autonomy and justice (Beauchamp and Childress, 1994). This has been applied to issues relating to our treatment of the food animals (Mepham, 1996).

Two key principles of ethics are utilitarianism and deontology. Utilitarianism is based on the principle of beneficence ('do good') and non-maleficence ('do no harm') to promote the wellbeing of the greatest number. This principle has obvious relevance to our approach to animals other than pets because we have little option other than to consider them *en masse*. Utilitarianism has become a rather discredited concept in human philosophy because, considered in isolation, it neglects the rights of the individual. In the context of our contract with animals it is not enough either. Philosophers describe our duty to respect individual rights as deontology, the science of duty, from the Greek *deon* (duty) and *logos* (science). It calls on us to respect the autonomy of all other individuals to those elements of wellbeing and freedom of choice that we might wish for ourselves, and is most simply expressed as 'do as you would be done by'. If we are to apply the practical principles of bottom-up ethics, we cannot label ourselves as either utilitarians or deontologists. We need to be both.

In recent years, it has become the custom to devise strategies for the promotion of health, safety and wellbeing in humans and other animals based on measures of outcomes, rather than provisions, as we measure the goodness of our actions by how well they work. This is a practical step in the right direction since outcomes are what matter. Measures for the principles of utilitarianism and deontology are, respectively, the wellbeing of the population and the autonomy of each individual. The two outcome measures of wellbeing and autonomy as applied to all parties determine the third outcome measure, which is justice. Justice demands that we, the moral agents, should seek a fair and humane compromise between what we take to meet our needs for food, sport, companionship, health and safety and scientific understanding, and what we give in terms of good husbandry, defined by competent, humane and sustainable action, for all life that comes within our dominion. This is not the same thing as giving all parties equal rights and will always fall short of the aspiration to seek total justice.

The 'just' approach to our treatment of animals is an unavoidably distorted expression of Thomas Hobbes view of the Social Contract where individuals consent to surrender some of their freedoms and submit to an authority in exchange for protection of remaining rights and the maintenance of social order (Hobbes, 1651). In the early days of domestication, it may be argued that wolves consented to give up some of their freedoms in order to derive the benefits of living among humans. It would be ridiculous to suggest that the farmed animals consented to becoming food. In my opinion it is valid to argue that, on good farms with good husbandry, quality of life for the animals (while it lasts) (as measured in terms of sustenance, comfort, companionship and security) can be as good as, or even better than, life in the wild.

Death and Killing

No discussion of the ethical issues involved in our interactions with the other sentient animals can avoid the subject of death and killing. First, a disclaimer. I have killed animals, as humanely as possible, in the course of my work. When sailing, I have caught just enough mackerel for lunch. Since the age of ten, I have never killed a mammal, bird or fish for sport. At that age, on a friend's farm, I shot a sparrow with an air gun, watched it die, thought 'what a pointless waste of a life' and decided there and then never to do it again. My views on the killing of animals are inevitably personal and I accept that many will disagree (in both directions). All I can say is that I have been living with this topic for many years and what I write now is the best I can do to achieve some sort of compromise between my conscience and reality.

To begin with, a few simple and incontestable statements of the obvious. Death is a fact of life for every living thing. Population control, by one means or another, is essential to the sustained life of the environment. The most worthy and organic gardener has to admit that 90% of their work involves the floral equivalent of butchery and slaughter. Being dead is not a welfare problem for the animal that has died, although it can be a problem for any offspring not yet mature enough for independent life. Unless death is immediate and totally unexpected, the act of dying does present problems of pain and fear. The claim that Shekita slaughter is humane because it is painless does not address the stress of fear in an animal conscious of choking to death in its own blood. The distress associated with dying and the approach of death will be a function of the intensity and duration of the period of suffering that precedes death. Whether the killer is a human, another animal, disease, or old age is irrelevant to the dying animal. A deer that gets a lethal shot from an expert stalker will suffer less than one hunted to exhaustion by humans or wolves and, in the latter case, being ripped to pieces. When a lion, that has depended on killing animals throughout adult life gets too old and toothless to hunt it is likely to suffer a prolonged decline before death from starvation or, when in extremis, being ripped to pieces by hyenas. The more difficult question is 'do (non-human) sentient animals fear the prospect of death?' We really don't know.

Sentient animals are clearly aware of danger when in proximity to a predator or in an area where they fear a predator may lurk. We may interpret this as fear of death, although we have no evidence that would enable us to conclude that they rationalize things this far. Because animals with sentient minds and the power of perception do not live only in the present, they remember sites and situations perceived to be dangerous and try to avoid them in the future. They also warn any dependent offspring involved against danger. Wild animals learn to live with threat and (another glimpse of the obvious) they only die once. Animals that perceive themselves to be under threat seek to maintain a safe flight distance from potential predators (whether humans or lions). Those forced to attempt an escape may be killed. Those that escape without harm remember that escape is possible so long as they are careful, so adapt to living under threat. Some years ago, I was examining records of fox hunting with hounds and discovered that the ratio of foxes roused by the hounds to foxes killed in the open field was thirteen to one. I think it reasonable to assume that the fox that escapes pursuit by predators develops the expectation that it can do it again, so does not live in a state of constant fear. Animals that manage to escape and survive having been badly wounded will, of course, suffer.

The existential fear of death as the end of existence is, it would seem, part of human nature. Unless one believes in the fires of hell, this is an irrational fear since being dead is not a problem, although it does, of course, mess up one's long-term plans. Fear of suffering during the process of dying is real. I am with Woody Allen on this one: 'It's not that I'm afraid to dic. I just don't want to be there when it happens'. Some animals, primates, cetaceans, elephants and rooks, appear to show signs of distress when faced with the death of a companion. This behaviour could reflect a sense of personal loss, compassion for a fallen comrade, or both, but I have no reason to suspect that their distress incorporates a sense of dread that at some undetermined time in the future this will happen to them. Because the death of all animals is inevitable, whether or not humans are directly involved, I have no fundamental objection to the killing of animals for food or for population control as an instrument of environmental management when the alternatives can be worse. I am in total agreement with those who argue that humans eat too much food of animal origin for the good of our health and that of the environment. I try, but oft times fail, to abide by the principle 'Eat food, not too much, mostly plants'. I also abhor several approaches to the management of 'natural' habitats, such as the slaughter of raptor birds so that humans can slaughter more grouse. Nevertheless, some natural environments such as the steppes of central Asia, the pampas of South America and the Taiga of the sub-arctic are sustained by stable populations of grazing animals. We cannot digest grass and we need to control the animal population. In these circumstances, not eating the meat from these animals would be ecologically spendthrift.

Because the killing of animals is necessary we have an obligation to ensure that the process of killing and everything that precedes that killing, from the first moment of disruption to the normal routines of their life to the point of insensibility, should be as quick and as humane as possible. This principle assumes the greatest importance in the case of the food animals (from the farm or from the sea), because of the numbers involved and because of the potential to cause suffering during harvesting, transport and lairage in the abattoir. This is not the place to go into detail on welfare issues relating to the transport and slaughter of the food animals. Many, notably the Humane Slaughter Association (the sister charity of the Universities Federation for Animal Welfare (UFAW)) have had much to say on this elsewhere. In the present context, the single message must be to strive to ensure that the animal to be killed passes from existence to non-existence with the absolute minimum of fear and distress. This principle should apply equally to the pig and to the lobster. Since death comes to us all, it may not be too much to suggest that this accords with the deontological principle of 'do as you would be done by'.

The killing of wild animals is an acceptable means of population control in order to preserve the balance of nature when the alternatives are worse, both for nature and for individual animal welfare. About 30 years ago, as a member of the UK Farm Animal Welfare Council, I visited the Oostvaardersplassen, which was established as a nature reserve on a reclaimed polder in Holland for birds that could come and go, as well as for populations of horses, deer and cattle, who couldn't. The plan was to leave nature to itself to establish an equilibrium. We visited in March at the end of a hard winter when the grazing animals were dying of starvation in their hundreds. The cattle, constrained by the design of their mouths to harvest long grass, were the most affected. The deer and horses, more able to browse and tear out the remaining grass by its roots, took longer to die but were wrecking the pasture, shrubs and trees. I have no hesitation in stating that this was the worst systematic abuse of animal welfare I have ever seen. The tragedy was that it was done with the best of intentions. The practice continued for many years until the winter of 2016/2017 when there were 3950 recorded deaths of cattle, deer and horses. Popular outcry finally forced a policy change towards population control through judicious culling.

Farms, Farmed Animals and Food

Table 3.1 applies the ethical matrix to the business of farming animals. In this model, the farmer is not just a food producer but also a steward of the living environment.

There are two moral agents in the matrix. The first are the farmers and landowners that are directly involved in the production of food from animals. The second is human society at large, which means any of us who derive any benefits from farmed land, whether for food, recreation or vital resources like water, soil and clean air, i.e. everybody. The moral patients are the farmed animals and the living environment.

The right of the human population to demand food that is wholesome, safe and fairly priced carries the responsibility to all other concerned parties, farmers, farmed animals and the environment. In practical terms, it means that the public must accept the need for legislation to ensure acceptable standards of animal welfare and environmental care. I suggest that we also have a moral responsibility to go beyond the constraints of legislation to encourage incentives to improve the husbandry of animals and land. The most effective way to achieve this has been through consumer action, e.g. paying more (when

Table 3.1. Food and farming: Moral agents and moral patients. (From Webster, 2022, with permission of John Wiley & Sons.)

	Beneficence	Autonomy	Justice
Moral agents Farmers Human society	Financial reward Pride in work Wholesome, safe, affordable food. Right to roam	Free competition Freedom of choice	Fair trade Support for good animal and planet husbandry Pay more for good husbandry
Moral patients Farmed animals	Competent and humane husbandry Conservation Sustainability	Freedom of choice Biodiversity 'live and let live'	'A life worth living' Respect for environment and stewards

we can) for food that carries the added value of local provenance, proven high welfare and/or organic production methods.

Justice for the producer balances the right to free trade and a decent income against the responsibility to practice good husbandry. According to the classic Adam Smith free-market argument (Smith, 1778), farmers should have no special rights. They are just one group within the overall division of labour so should be served no better or worse than any other group by the 'invisible hand of the market'. If the sole function of the farmer was to provide food, this argument would apply. However, farmers also carry the direct responsibility of sustaining the living environment, not just for us but for the sake of all life. This is a long-term commitment that cannot be addressed through the short-term economics of the free market. It is the responsibility of society at large. What this means in practice is that public money should be spent on public goods. Ideally, individual members of the public should meet the full cost of private goods, which implies that there should be no subsidies on food production per se. Individual buyers should be free to select the importance they give to price and production standards, which implies that those without money-worries should not have the right to impose high prices for value-added foods on those who cannot afford them. Stewardship of the environment (planet husbandry) is a public good, so should be supported entirely from public funds (i.e.taxation). When I first wrote this in 2021, the UK government were examining ways to redirect agricultural subsidies towards the principle of public money for public goods. Since then, they have taken a few uncertain steps forward but, as I write now, may be in retreat.

Justice for the moral patients requires us to provide the farmed animals with a life worth living. We can promote the wellbeing of the flock or herd through competent and humane husbandry. This may be expressed simply in terms of the Five Freedoms (FFs), i.e. the freedom from hunger and thirst; chronic discomfort; pain, injury and disease; fear and stress; and the freedom of choice. The first four freedoms address the utilitarian principle of beneficence. Freedom of choice addresses our responsibility to recognize the autonomy of the individual by providing opportunities for the animals to make a constructive contribution to their own quality of life. This may involve selection of diet, environment (e.g. indoors or outdoors) and social expression. The utilitarian approach to the management of the living environment requires that we should strive to sustain the quality of the habitat and conserve all life within it. The principle of deontology identifies our responsibility to respect the individual needs of all fauna and flora, whether or not they make any contribution to our own wellbeing. We need to be realistic about this. We cannot, and should not, seek to preserve every rat and every tree. Nevertheless, we have a duty to respect the needs of all lifeforms, which implies that we should always seek to minimize individual suffering and lasting harm. This requires a policy of critical, but constructive, support to the farmers and landowners directly responsible for putting our principles of respect into practice.

Animals in Laboratories

Humans maintain and carry out experiments on animals in laboratories for the advance of knowledge through science, as well as for the protection of health and safety by testing a wide range of chemicals, such as medicines, garden sprays and common household products for evidence of toxicity. This is a utilitarian pursuit for the general wellbeing of the population of humans, domesticated animals and plants of nutritional or cultural interest. As with the farmed animals, the ethical issues are confounded by the fact that in the cost: benefit analysis all the benefits accrue to us, the moral agents, and all the costs are borne by them, the moral patients.

There has been much progress in recent years towards reducing the cost to animals of scientific procedures considered necessary for human health and safety. One big step in this direction has been adoption of the principle of the 'Three R's', reduction, replacement and refinement. This calls on us to reduce the number of animals involved in experiments where possible, replace animal-based tests with *in vitro* (test tube) experiments and, when it is essential to use live animals, refine the procedures to minimize the risk of incurring suffering. A big step forward was the UK Animals (Scientific Procedures) Act 1986, which requires that procedures involving laboratory animals should only be allowed subject to a harm: benefit analysis where the cost to the animal in terms of 'pain, suffering, distress, or lasting harm' can be justified in terms of the potential benefit to humans (or other animals). The more recent EU Directive (2010/63/EU) carries a similar message. There can be no doubt that these directives have greatly improved conditions for laboratory animals, not least because they compel those who obtain their livelihood from working with these animals, whether directly or at a distance, to give more thought to how the animals might feel.

Table 3.2 uses the structure of the ethical matrix to examine the needs of the moral patients and the needs and responsibilities of the moral agents in the matter of procedures with laboratory animals.

The first of the listed moral agents is human society at large, i.e. any of us who depend on these procedures for our health and safety. We, as individuals, have the right to select from a range of products according to our individual perception of their efficacy and the possible cost of their use to laboratory animals and the environment. This freedom gives us the responsibility to develop a compassionate and informed understanding of what it is that we are buying. The regulators of scientific procedures have to balance their responsibility to society and to the experimental animals. In accordance with Regulation (EC) No 1223/2009 of the European Parliament, the use of experimental animals for the testing of cosmetic products has been illegal since 1998 in the UK and since 2009 in the EU (Regulation (EC) No 1223/2009; Government UK (1999). The principle of beneficence determines that animal welfare be enshrined in Codes of Practice. Autonomy, along with the interests of animal welfare, require the regulators to be open-minded and sympathetic to new ideas. This is particularly relevant to the business of routine toxicology testing, where it is too easy to stick with old routines when more refined, less stressful alternatives become available because 'it has always been done this way'.

Developers and producers of new drugs, household goods and garden sprays have the right to financial rewards for their work and to free competition, unimpeded by restrictive legislation. In return, they have the responsibility to promote the welfare of both the experimental animals and the staff with direct responsibility for their care. The animal care staff in scientific laboratories carry out physically and emotionally taxing work with a great deal of skill and compassion. It is the responsibility of their bosses to ensure not only that they can take pride in their work but also that their voices are heard with regard to day-to-day and strategic decisions in relation to matters of animal husbandry and welfare.

Our first duty to the animals is to minimize harms arising from the procedures themselves through just application of the harm: benefit analysis. In the UK, this is enshrined in law and effectively policed through formal inspections and the presence of a named veterinarian whose first responsibility is to the animals, not the procedures. However, seeking to minimize harm during the procedures is not enough. Most animals in laboratories, most of the time, are not actually undergoing potentially harmful procedures but may be living a depressing life in barren cages. When not constrained by the needs of a specific procedure, all these animals should be given the opportunity for as much environmental and social enrichment as possible.

Generally speaking, laboratory animals receive more protection from the law than farmed animals and are less likely to suffer neglect. It is, moreover, instructive to compare the relative magnitude of welfare problems for farmed and laboratory animals. In my book, *Animal Welfare: A Cool Eye towards Eden* (Webster, 1994), I wrote: 'In the UK, the average human omnivore who

Table 3.2. The ethical matrix: laboratory animals. (From Webster, 2022, with permission of John Wiley & Sons.)

	Wellbeing	Autonomy	Justice
Moral agents			
Society	Health and safety	Choice	Transparency
Regulators	Welfare standards		Codes of practice
Managers	Welfare standards		Apply the three R's
Staff	Pride in work	Free competition	Wellbeing
	Animal care	'a voice'	Staff and animals
Moral patients			
Lab animals	Minimizes harm	Enriched environments	Compassionate Interpretation of standards

maintains a good appetite to the age of seventy will consume 550 poultry, 36 pigs, 36 sheep and six oxen. The number of mice sacrificed to advance knowledge and improve human health and safety is two.' I suggested in 1994 that this may not be too much to ask of brother mouse. The most recent statistics from the Home Office (2023) show that this number has not significantly changed.

Wild Animals in Captivity

I suggest that our best policy with regard to animals in the wild is to leave them well and leave them alone. To achieve a sense of wellbeing, they need an environment that provides the full range of the resources they require to meet their physical and emotional needs during all seasons. For the larger mammals, this implies a varied habitat and a lot of space. Currently, this space is being eroded at a rate that poses an existential threat to many species. The most effective way to address this problem is through the establishment of large nature reserves. When this is not possible, smaller reserves can be strategically linked by wildlife corridors that allow free movement of animals between them. Since it is human nature to invade and erode these areas of wilderness for our own devices, it will nearly always be necessary to manage and police these reserves to prevent poaching and destruction of habitat. This costs time and money, so it makes complete sense to provide financial support from humane, environmentally sustainable tourism. Management of even the most extensive of these reserves is likely, on occasions, to require a controlled cull of iconic, but environmentally destructive, species like elephants in order to avoid a catastrophic failure of good intentions such as the Oostvaardersplassen fiasco.

The aim of nature reserves is to sustain a habitat wherein wild animals adapted to that habitat have the freedom to live the life for which they are best suited and, if these are managed well, this would seem to be beyond reproach. More contentious is the business of keeping wild animals in captivity for the enlightenment and entertainment of the paying public. When I was a small boy living in Bristol, I greatly enjoyed trips to the zoo to get a ride on the elephant and gaze in wonder at large cats (and the famous gorilla, Alfred,) in cages. In common with almost everybody else at the time, I gave little thought as to how the animals might feel about life in these conditions and I was totally unaware that most of them had been captured from the wild and locked up for life. Thankfully, zoos like this are no longer tolerated in most societies. Much thought has been given to the creation of enriched environments that seek to reproduce the resources of the natural habitat within the confines of available space and the need to give the public chance to see the animals.

Most zoo animals are now bred in captivity or, when at threat of extinction in the wild, brought to zoos and wildlife parks for breeding to preserve the species and, in a few cases, for return to the wild. I shall not dwell on the motivation of those who run zoos for the public, not least because this book is about animals, not people. I would just say that my experience of those who work in zoos at all levels is that they love their animals and most of them understand them pretty well. In keeping with my general theme, my question is 'How do 'wild' animals feel about life in captivity?' I put the word 'wild' in quotation marks because the phrase wild animal, like farm animal, is an anthropocentric conceit that classifies animals in terms of their utility to us.

When we are unsure about how to act in relation to an animal of which we know little, it is best to start from first principles. Any sentient animal, however we may choose to categorise it, is driven by the motivation to satisfy its physical and emotional needs and seek the freedom to control its own quality of life. Basic needs include freedom from hunger and thirst, pain, fear, injury and disease, and the freedom to seek comfort and security. Advanced psychological needs at the deeper levels of the sentient mind may include freedom to express curiosity and pleasure, play and have a good social life. The relative importance of these different needs will differ between different species adapted over generations to domestication or life in the wild. However, these principles can be applied to any set of questions relating to the basic and advanced needs of sentient animals, wherever they may be, addressed to those with direct responsibility for control of their lives and answered on a species-by-species basis by those with the competence to judge.

I illustrate this principle with two examples. The need to eat is the strongest of the life forces and one that cannot be measured simply in terms of the physical need to maintain normal body function through the provision of adequate nutrition. Sentient animals have a strong emotional need to forage or hunt for food and derive great satisfaction when that need is met. Good zookeepers know well that their animals derive satisfaction from being made to work for their food reward and will go as far as reasonably practical to make the hunt as realistic as possible. Obviously, offering live goats to tigers is not on, although the ban on live prey is not universal. For example, some snakes may only eat if presented with a live mouse. With herbivores, the situation is rather different. Equines, by virtue of the design of their digestive tract (small, simple stomachs and fermentation that occurs in the hind gut) have a physical need to forage for many hours. Elephants in the wild, by virtue of their large appetite for foods of low digestibility, need to travel for long distances to meet their nutrient requirements. Some argue from this observation of natural behaviour that elephants have a basic need to take long walks in wide open spaces. Possibly they do, but I suggest that it is more likely that they are motivated to spend many hours foraging for food, rather than simply going for a long walk. With both carnivores and herbivores, the aim should be to make the search for food a pleasant and rewarding way of passing the time.

Meeting the needs of animals in captivity for a secure and satisfactory social life is a critical issue that must be addressed on a species-by-species basis and by those with the competence and experience to speak for the species in question. Some species crave social contact and suffer in isolation, whilst others are adapted to a solitary existence. Others still can elect to form close relationships with individuals, not necessarily of the same species, but remain permanently antagonistic to others. Some animals experience fear and stress when humans are in sight and they perceive they have no place to hide. Others clearly enjoy human company. In captivity, these will be learned choices rather than a hard-wired property of the species. I observed and photographed a young rhino in the extensive paddocks of Longleat Safari Park approach one of the animal attendants (a veterinary student with a summer job) to get a tickle under the chin. This could hardly be called 'natural' rhino behaviour but, given freedom of choice, this young rhino had elected to seek out something that it had learnt would give it pleasure.

Animals in Sport and Entertainment

Throughout history, animals have been used by humans for a wide range of entertainments, most of them barbaric. Lethal pursuits included the baiting of bulls and bears, bull fights, dog fights, cock fights and eating Christians. Non-lethal attractions for our titillation have involved performing animals, especially in circuses. The animal most commonly required to perform for humans in the pursuit of sport and recreation is the horse, exploited for its athletic prowess and adaptability to human training methods. Performances that reflect agricultural practices and traditions include rodeo and sheepdog trials. I exclude from this discussion any consideration as to the morality of legal sporting pursuits that involve the killing of animals (hunting, shooting, fishing) on the grounds that this book is an enquiry into the minds of sentient animals and no sentient animal in its right mind would voluntarily wish to take part.

The three questions that need to be addressed are:

- Can the performance, whether in the circus or on the racetrack be a source of suffering?
- Do the training methods deemed necessary to prepare the animals for the performance involve the deliberate imposition of pain or fear?
- Do the living conditions of the animals meet their needs for the times when they are not performing? (i.e. most of the time.)

Horse racing exploits the strong motivation of the horse to run, and to run in company. In nature, and when not driven by jockeys, some horses prefer to lead whilst others prefer to follow. In general, I believe that horse racing, whether on the flat or over fences, is an activity that depends on the willing, indeed enthusiastic, participation of the horses. Horses do, of course, get serious and lethal injuries on the racetrack and these are of great concern to owners, carers and the general public. Those with the most severe injuries are likely to be killed on the track, so they don't suffer for long. Those with less severe injuries that are allowed to recover may never race again so will not experience the fear of its recurrence. In this regard, racing differs from eventing when horses are expected to clear extremely challenging jumps, frequently come to grief but are expected to go through the same routine again and again. I once heard a famous event rider talk of 'Badminton virgins', i.e. young enthusiastic horses with, at that stage of their lives, no experience of falls and therefore no fear. He also spoke of experienced horses that learned to be more circumspect and of occasions when he has pulled up a horse because it was clearly hating every moment of the ride. This very experienced and sympathetic rider was making the point that eventing can become distressing to some horses.

The biggest point of contention in regard to the welfare of racehorses relates to the use of the whip. This is, by definition, based on negative reinforcement, the stimulus of pain (or fear of pain when the jockey shows the horse the whip) to drive the animal harder. There are strict rules on the design and use of the whip intended to minimize stress, but it could never be construed as other than a violent act carried out exclusively for the benefit of the humans involved (jockey, owner and trainer). There is now pressure to ban the whip altogether. Before this proposal is dismissed as sentimental nonsense, we should consider, firstly, whether a whipped horse actually runs faster but the evidence for this is shaky. Secondly, we should consider whether racing without whips would make any difference to the excitement generated by the race and the attractions of betting. Horses are going to finish the race in an order determined by their fitness, their motivation and the skill of their jockey. This will happen whether or not the jockeys carry whips and the bookies and punters will, on average, be neither better nor worse off. The owners may grumble.

Paradoxically, the equestrian event most open to informed criticism on welfare grounds is the one that involves the least effort and the least risk, namely dressage. The horses are trained to perform a number of unnatural manoeuvres while their head and neck are forced into an unnatural posture. To achieve this unnatural posture many dressage horses are compelled to be held on a tight rein pulling on bridles with an extremely tight nose band. This forces the horse to adopt the unnatural head position because it is painful to do otherwise. Moreover, there is evidence that the prolonged use of tight nose bands can lead to lasting damage to the mouth. Once again, the traditionalists argue that dressage, as currently practised, is a supreme demonstration of the ability of the horse to perform a complex series of manoeuvres devised by us. It is far more impressive to watch a demonstration of the skills of horse and rider when this is done without recourse to bit and bridle. A great example of this was demonstrated by Stacy Westfall's winning performance at the Freestyle Reining competition during the All American Quarter Horse Congress in 2006. (The video can be viewed on YouTube at https://www.youtube.com/watch?v=TKK7AXLOUNo).

The circus has been the traditional place to watch performing animals. Once again, during my boyhood days, these animals included not only horses, but also lions, tigers and seals. Today, the use of animals such as the large cats, which had to be trained by methods based on fear, is (entirely?) extinct. The few animal acts we see in the circus today usually involve animals like dogs and horses, who can be trained by positive reinforcement. When these animals are trained well and treated well, I do not think they present a welfare problem. Some will argue that dressing animals up in silly costumes is an insult to their dignity. If we stick strictly to our understanding of animal behaviour, this will only present a problem for animals that possess a sense of self and non-self, and therefore an awareness of what they look like. We may exclude horses from this category. There is some experimental evidence that dogs have a sense of self and non-self so could be distressed by the fact that their appearance is unnatural. It is much more likely that dogs pick up human signals that cause them distress because they sense they are being laughed at. They will obviously be irritated if the article of apparel itches or interferes with their behaviour. I think it entirely proper to object to human actions that abuse the dignity (the telos) of our fellow mortals on the grounds that they reflect some of our most extreme displays of sentimentality and anthropomorphism. However, I say again, as far as the animals are concerned, it is not what we think but what we do that matters. It follows that I am not too concerned by performing animals acts that cause no distress to the animals, either in performance or in training. My main concern as to the welfare of animals in circuses (and bad zoos) relates to question three. The quality of their living accommodation can fall far short of that necessary to meet the FFs.

I include in my list of animal performances rodeos and sheepdog trials. Many of the events in a western rodeo are designed to show off skills necessary to the work of the traditional cowboy, but two of the most spectacular events, bronco riding and bull riding, are completely pointless. The rider has to spur the bucking animal and stay on board for 8 seconds. Equal points for performance are awarded to animal and rider. This looks like a violent and cruel practice, although the only animal likely to be injured is the cowboy. Aversion techniques are used in training animals for the rodeo but, once trained, they become seasoned performers. Moreover, the contest, as seen through their eyes, is one they always win, since the cowboy always comes off after no more than 8 seconds. I have spent some time observing the behaviour of experienced rodeo horses and bulls and, generally speaking, out of the ring, they appear to be relaxed and stress-free. It is a useful exercise in ethical analysis to compare this with the experience of sheep in that most serene and British of pastimes, the sheepdog trial. When dog and handler are experts, the sheep will be moved with minimal distress at a controlled speed and in the right direction because all parties – sheep, sheepdog and shepherd – understand the rules of flight distance. When things go wrong, the sheep run off, maybe not in panic but instead under stress. In any event, it is a game that the sheep will always lose. I am a fan of sheepdog trials because I believe they are a beautiful display of human and dog working in combination to perform a task essential to the humane and sensitive management of an only slightly domesticated animal. Moreover, when viewed through the eyes of the animals, I cannot see much wrong with much of rodeo practices either. However, there is one exception. The only rodeo practice that has any real relevance to the practical skills needed by the cowboy is the one most likely to cause distress, namely calf roping, which involves the lassoing a young calf from horseback.

Pets

We cannot escape the fact that our pets (with the exception of cats) are the animals whose lives are most affected by our behaviour. As dogs are almost entirely dependent on us, they are potentially most at risk from abuse. While starvation and neglect may be rare, emotional stress, such as separation anxiety, is all too common. Horses that evolved to graze the open plains in a herd of close compatriots can suffer physical and emotional stress as a result of improper feeding and social isolation. Breeding animals to suit the whims of fashion abuse our responsibility to ensure the fitness of animals in our care. For reasons of fashion, or perceived cuteness, we breed dogs that are prone and, in some cases, condemned to suffer from conditions that are entirely of our own making. These include chronic respiratory distress and the need for repeat caesarean sections. Even horses are now being bred for traits that may look pretty (or fashionable) to our eyes but impair their normal fitness and athleticism. I cite the breeding of Arab horses with noses dished to a degree that significantly impairs respiration, especially during exercise.

We expect some animals that are classified by us as pet or companion to do more than conform to a lifestyle of our choosing. We also expect them to do our will. This applies most obviously to dogs and horses. When this is done well it can be satisfactory for both parties, especially in the case of dogs whose minds are naturally inclined to seek the approval of their leaders and make a positive contribution to the welfare of the extended family. Well-trained contributors to the general good, such as police dogs, guide dogs and sniffer dogs undoubt- edly derive satisfaction from their work and live a more fulfilling life than those shut up indoors with nothing to do all day. Well-trained dogs and horses are emotionally stable and their quality of life is likely to be more than just satisfactory. Expert trainers, consciously or otherwise, understand the minds of their animals. The sentient minds of non-human animals are complex but not complicated. They recognize direct, consistent, unambiguous signals and learn to react as we would wish when correct responses are rewarded by posi- tive reinforcement. Domesticated animals cannot be expected to do our will if they fail to understand the message, the messages are inconsistent, or they are punished for doing what they imagine to be the right thing.

I recognize that most of the people who insult the natural integrity or confound the minds of their pet animals are unaware that they are doing any harm. Should any of them read this book and are anxious that they may fall within this category, I would ask them to follow the simple message that applies to all who have responsibility for the care of sentient animals, whether in the home or on the farm, in laboratories or zoos. Get as much good advice as you can from those with real experience, then work from first principles. Lay out the set of basic and advanced physical and emotional needs that apply to all sentient animals. Here, the FFs can provide a sturdy framework on which to hang the details. Apply these basic principles to meet the special needs of the animals for whom you have responsibility within the environment you have imposed on them. Regularly review the outcome of your actions in the context of how well the physical and emotional state of your animals appears to meet these FFs and, if necessary, modify your own actions. Your aim must be to ensure, as far as possible, that you and the animals in your care are of the same mind.

Note

[1] This letter has been largely drawn from Chapter 11 'Sentience and the Sentient Mind' in *Animal Welfare: Understanding Sentient Minds and Why it Matters* (Webster, 2022).

References

Animals (Scientific Procedures) Act (1986). Available at: https://www.legislation.gov.uk/ukpga/1986/14/contents

Beauchamp, T.L. and Childress, J.F. (1994) *Principles of Biomedical Ethics*. Oxford University Press, New York.

Brabazon, J. (1995) *Albert Schweitzer: A Biography*. Syracuse University Press, New York.

EU Directive (2010/63/EU) Protection of Animals used for Scientific Purposes. Available at: http://data.europa.eu/eli/dir/2010/63/oj

Government UK (1999) Guidance on the Regulations as they Apply to Cosmetic Products Being Supplied in or into Great Britain.

Hobbes, T. (1651) *Leviathan*. Penguin Classics, London, UK.

Mepham, B. (1996) Ethical analysis of food technologies: An evaluative framework. In: Mepham, B. (ed.) *Food Ethics*. Routledge, pp. 101–119.

Regulation (EC) No 1223/2009 of the European Parliament and of the Council of 30 November 2009 on Cosmetic Products. Available at: http://data.europa.eu/eli/reg/2009/1223/oj

Smith, A. (1778) *The Wealth of Nations*.

Webster, J. (1994) *Animal Welfare: A Cool Eye towards Eden*. Blackwell Science, Oxford.

Webster, J. (2022) *Animal Welfare: Understanding Sentient Minds and Why it Matters*, UFAW Animal Welfare Series. John Wiley & Sons, Oxford.

YouTube (2006) Stacy Westfall Championship Bareback & Bridleless Freestyle Reining with Roxy. Available at: https://www.youtube.com/watch?v=TKK7AXLOUNo

Freedoms and Domains

4

The welfare of any sentient animal is determined by its own perception of its physical and emotional state. This applies equally to the huge population of food animals as it does to the pets on whom we may lavish individual attention. Increasing public concern for action to improve animal welfare has generated the demand for animal welfare science that seeks to improve our understanding of the nature of animal emotions and motivation and, from this, improve the quality of our care. The animal welfare scientist has a responsibility not only to do research and publish papers to be read by other scientists, but also to communicate new knowledge and understanding in a manner that is appropriate to the full spectrum of individuals in society, including fellow scientists; those directly involved in the care of animals on farms, in laboratories, zoos and in the home; and those who may have little direct contact with animals but derive from them some utility or pleasure, i.e. everybody else.

Animal welfare science is a big topic, since it embraces everything that may affect the physical and emotional state of the animal, its ability to cope and its quality of life. If we are to attempt a comprehensive analysis of the challenges to the welfare of a sentient animal and the consequences for its quality of life, we need some ground rules. In 2016, David Mellor introduced the concept of the Five Domains (FDs), a refinement of the Five Freedoms (FFs) (FAWC, 1993), to serve as a framework for overall assessment of quality of life in animals. The FDs are made up of four input categories: nutrition, environment and health, which are categorized as survival-related factors, and behaviour, which is classified as a situation-related factor but might better be described as opportunity to express rewarding behaviour, since behaviour itself is an outcome. The fifth domain is mental state, the outcome for the animal expressed in terms of negative and positive experiences. It is this domain that determines its physical and emotional state, i.e. its welfare status.

David Mellor presented the FDs as an alternative and a successor to the FFs 'in the light of new scientific knowledge and understanding of animal welfare' (Mellor, 2016). The editors of the journal *Animals* invited me, as the original proponent of the FFs in their current form, to contribute an opinion piece to complement his paper and, in essence, to discuss the relative validity and utility of the two approaches. My immediate, short answer is that it depends on who you are addressing: scientists, legislators, animal workers, or the general public. However, I need to explain this further.

The Five Freedoms

History

This phrase began life as the four freedoms, introduced by Franklin Roosevelt in his address to the US Congress in 1941. He identified these four freedoms as freedom of speech, freedom of worship, freedom from want and freedom from fear. It should be obvious that these, like the later FFs, are aspirations. He was not making it an article of law that all the people should experience all of these perfect freedoms all the time. They are, however, memorable.

The phrase was commandeered in 1965 by the Brambell Committee report on the welfare of farm animals in intensive systems to summarize their conclusion that farm animals in confinement should be allowed sufficient space to permit the following five minimal behaviours or activities, to stand up, lie down, turn round, stretch their limbs and groom all parts of the body (Brambell, 1965). When I joined the UK Farm Animal Advisory Committee (the predecessor of the FAWC) in 1979, I suggested that while these things were of vital importance to animals in the most intensive systems, they presented a very restricted view of farm animal welfare and left many, indeed most, welfare problems off the page. Invited to come up with something more comprehensive I first proposed a new set of FFs. The FAWC worked on this original set of freedoms and, in 1993, they published an updated version that matched each of the FFs with five provisions:

1. Freedom from thirst, hunger and malnutrition by ready access to a diet to maintain full health and vigour;
2. Freedom from thermal and physical discomfort by providing a suitable environment, including shelter and a comfortable resting area;
3. Freedom from pain, injury and disease by prevention or rapid diagnosis and treatment;
4. Freedom from fear and distress by providing sufficient space, proper facilities and the company of the animal's own kind; and
5. Freedom to express normal behaviour by ensuring conditions that avoid mental suffering.

The alert will spot that these 'five freedoms' include, in fact, eleven. In recent years, I have preferred to describe the fifth freedom as freedom of choice. In practice, this, like all the rest, has to be a partial freedom. These freedoms are aspirations, not directives. All are outcome measures. The provisions outline the husbandry necessary to promote these outcomes.

The pan-European Welfare Quality® assessment protocols for the welfare of farm animals identify four welfare principles defined by twelve criteria. The

FDs recognize 15 or 18 'negative affects', depending on how you count them (Welfare Quality⸱: Assessment Protocols for Cattle (etc.), 2009). Other publications have produced much longer lists. This is a process analogous to inflating the Ten Commandments into the *Book of Leviticus*: worthy but very dull.

My case for the FFs has always been that, at a very simple and basic level, they are comprehensive. Moreover, attempts to strengthen the case by adding detail can have the opposite effect since the more one tries to expand the argument by adding examples, the more likely one is to leave things out. This makes a very important point, especially in the context of legislation. The two key pieces of animal welfare legislation in the UK have an elegant simplicity. The second definition of cruelty in the UK Protection of Animals Act 1911 states 'to cause unnecessary suffering by doing or omitting to do any act'. The recent UK Animal Welfare Act 2006 goes further by introducing a 'duty of care' not only to avoid conditions that may lead to suffering, but also to promote positive welfare. The strength of both these Acts, in my opinion, is that they stick to first, and timeless, principles.

Limitations and strengths

Mellor proposed several limitations of the FFs. I have some more of my own. His strongest criticism is that, unlike the FDs, they do not embrace the concept of positive welfare. I concede this point and shall return to it later. I also concede that as a practical series of recommendations for good husbandry, the five provisions, or the four welfare principles defined by Welfare Quality⸱ may be of more direct practical use as a guide to humane husbandry than a simple statement of freedoms. However, I believe that the great strength of the FFs in practice is that they describe outcome indicators. In recent years, outcome measures have become the standard approach to the development of quality control protocols, whether for the evaluation of animal welfare on individual farms or within overall production systems such as RSPCA Assured, (formerly Freedom Foods). I also concede that the FFs do not capture 'the breadth and depth of current knowledge of the biological processes that are germane to understanding animal welfare and to guiding its management', but that was never their intention. I suggest we are getting into Leviticus territory here.

There are two ways in which I believe that the FFs fall short. The first is that they only describe a snapshot or an attempt to define welfare at a moment in time. They do not properly reflect the causes and consequences of stresses that lead to long-term problems, e.g. behavioural problems, such as learned helplessness due to long-term denial of normal behaviour in sows, or physical problems, such as metabolic exhaustion in the dairy cow. The message here is that any outcome-based working protocol for the evaluation of animal welfare must include chronic indices of failure to cope with physical and emotional challenge.

My other concern relates to the fifth freedom 'to express normal behaviour'. This is the only freedom **to**, all others are freedoms **from** and, I believe, beyond cavil. Freedom **to** begs the question of what is normal behaviour? Does it include complete sexual freedom? Clearly not. Does it include freedom of one individual to compromise the welfare of another? I hope not. My freedom to swing my fist should stop at the point of your nose. The definition of normal behaviour can be interpreted quite sensibly so long as it does not get too bogged down in sophistry. Nevertheless, in recent years I have come to believe that the fifth freedom would be more neatly expressed as 'Freedom of Choice'. This incorporates the freedom to express natural behaviour with regard to choice of diet, environment, social contact, comfort and security. Once again, the concept of freedom of choice needs to be interpreted responsibly. Animals, like children in our care, are our 'Moral patients' (*see Letter 3*) and should not, for example, be given free licence to eat themselves to death. I believe the concept of freedom of choice addresses all the concerns set out in the original Brambell report into the welfare of farm animals within intensive systems (being able to stand up, lie down, turn round, stretch their limbs and groom all parts of the body). More generally it addresses my greatest criticism of the business of factory farming, namely that it is more or less totally designed to promote their own quality of life at the expense of the animals.

The Five Domains: Strengths and Limitations

Briefly restated, the FDs approach firstly identifies four categories of input factor that act upon the mind and body of a sentient animal and then assesses their impact on a series of outcome indicators of mental state (15 negative and 13 positive effects). These may or may not then be integrated into an overall measure of welfare status. The individual outcome indicators provide a comprehensive structure upon which animal welfare scientists can build their knowledge and understanding of specific topics and identify topics for new research. They can also be used as the foundation for outcome-based protocols for the evaluation of animal welfare on farms, in zoos or research establishments and so on.

Integration of the elements of the fifth domain into a single measure of welfare status may be, in some circumstances, necessary, e.g. for the overall classification of an individual farm as acceptable/unacceptable, or to give it a score within a ranking system as proposed by Welfare Quality® or the '5-Step Animal Welfare Standards' developed in North America (Global Animal Partnership, 2008).

For presentation to the general public any quality assurance (QA) scheme will usually need to provide an overall score as to acceptability or quality ranking, despite the difficulties inherent in offsetting a good score in one category against a poor score in another. However, whatever the overall score (unless everything is perfect), I argue that, as far as the animals are concerned, the most important purpose of any welfare-monitoring scheme is to identify and address specific problems. In this regard FFs can be used to identify a comprehensive, specific, step by step, series of outcome indicators calling for action.

'A Life Worth Living'

The concept of Quality of Life (QoL) recognizes that animals have both positive and negative experiences and focuses on the balance between the two. The FAWC developed the notions of 'a life not worth living', 'a life worth living' and 'a good life' (FAWC, 2009). Green and Mellor (2011) formulated a four-tier QoL scaling system with two positive categories above and two negative categories below a neutral point of balance. Mellor suggests that this approach is 'more likely to be effective as a motivational framework than as an effective foundation for developing regulations'. I agree. I recognize the circumstances wherein an overall assessment of QoL can be of value, most obviously when a veterinarian is communicating with a pet owner that is faced with the prospect of euthanasia. I also recognize its utility as a basis for ranking farm overall animal welfare standards within QA schemes.

I do suggest, however, that there are circumstances where the QoL concept is not particularly helpful and may indeed be counter-productive. My first concern is with the suggestion that it is possible to define QoL as the algebraic sum of positive and negative experiences. In the case of the dairy cow, can one really quantify the extent to which, for example, 'affectionate sociability' can offset the pain of chronic lameness? More generally, we must acknowledge that our interpretation of the feelings of others can only be subjective. Since I can never be entirely sure how you are feeling, I am reluctant to speak with authority on the mental state of a dairy cow. In a strictly practical sense, one should, wherever possible, avoid the idea that a specific harm can be offset by another good. If there is a significant harm of any sort, efforts should be made to remedy it.

I also have my doubts about the concept of 'a life worth living' because it is a value judgement made by us, rather than by the animal in question. An insensitive farmer may consider that the life of a severely lame dairy cow has worth, so long as she continues to give milk. The highly sensitive owner of an infirm geriatric dog may consider that its life is worth prolonging because it continues to give and receive love. What these two extreme examples have in common is that in neither case does the animal contribute to the decision. The conclusion as to whether or not the life of a domestic farm or pet animal is worth living is something that we humans will make on behalf of the animal, based on how we think it feels when experiencing a physical and social environment largely dictated by us. A human teenager, exposed to such paternalism, or a family living under colonial rule, may be fully justified in protesting 'What right have you to tell me how I feel?' Thus, while I concede that QoL may be effective as a motivational framework, as far as the animals are concerned, it is not what we think or feel but what we do that counts.

Five Freedoms vs Five Domains: Who Needs Them?

The aim of this letter has been to discuss the relative validity and utility of the FFs and FDs – two approaches to the outline analysis of animal welfare. I hope by now I have made it clear that I believe the aims of the FFs and FDs are different but complementary. The FDs seek to assess the impact (specific and overall) of the physical and social environment on the mental (affective) state of a sentient animal. The FFs (with the five provisions) is intended as an outcome-based approach to identify and evaluate the efficacy of specific actions necessary to promote well-being. However, both have utility.

The FDs are clearly of use to animal behaviour and welfare scientists because they can embrace new knowledge and understanding, and provide pointers for new study. They can also be used for an in-depth analysis of the impact of specific management practices (human actions) on animal welfare. For example, the FD approach has recently been used to evaluate the negative (adverse) welfare impacts of a range of procedures to which domestic horses may be subject, across a broad range of different contexts of equine care and training (McGreevy *et al.*, 2018). This has been a valuable exercise. In the case of procedures that may be deemed necessary, such as castration, it encourages us to think carefully as to what constitutes both best practice and minimally acceptable practice. For other procedures, such as the use of the whip in horse racing, it addresses the question as to whether the alleged 'benefits' can ever justify the cost. In this, and in many other examples, the FD approach provides a highly effective foundation for research and evidence-based conclusions as to the impact of the things we do on the mental state of the animals in our care.

The FFs are much simpler (perhaps too simple for scientists), but are based on fundamental, timeless principles that do not need to be re-evaluated in the light of new research. They do not attempt to achieve an overall picture of mental state and welfare status. They are intended as no more than a memorable set of signposts to right action.

Since, as far as the animals are concerned, it is not what we think or feel but what we do that counts, I suggest that they are likely to have more impact on, and be of more use to, everybody else – that includes the animals.

Note

[1] This letter is based on Webster (2016), Animal Welfare: Freedoms, Dominions (sic, printer's error) and 'A life worth living'. *Animals* 6 (6) 35.

References

Brambell, F.W.R. (1965) *Report of the Technical Committee of Enquiry into the Welfare of Livestock Kept under Intensive Conditions.* HMSO, London, UK.

Farm Animal Welfare Council (FAWC) (1993) *Second Report on Priorities for Research and Development in Farm Animal Welfare*. DEFRA, London, UK.

Farm Animal Welfare Council (FAWC) (2009) *Farm Animal Welfare in Great Britain: Past, Present and Future*. FAWC, London, UK.

Global Animal Partnership (2008) *The Five-Step Programme*. Available at: www.global partnership.org

Green, T.C. and Mellor, D.J. (2011) Extending ideas about animal welfare assessment to include 'quality of life' and concepts. *New Zealand Veterinary Journal* 59, 316–324.

McGreevy, P., Berger, J., Brauwere, N., Doherty, O., Harrison, A. *et al.* (2018) Using the Five Domains Model to assess the adverse impacts of husbandry, veterinary and equitation interventions on horse welfare. *Animals* 8, 41.

Mellor, D.J. (2016) Updating animal welfare thinking: Moving beyond the 'five freedoms' to 'Life worth living'. *Animals* 6(3), 21.

Webster, J. (2016) Animal welfare: Freedoms, dominions and 'A life worth living'. *Animals* 6(6), 35.

Welfare Quality (2009) Assessment Protocols for Cattle, Pigs and Poultry; Welfare Quality Consortium: Lelystad, The Netherlands. Available at: https://www.welfarequalitynetwork.net/en-us/reports/assessment-protocols/

Welfare Quality Assurance: The Virtuous Bicycle

5

The 2002 Curry Commission Report on the Future of Farming and Food saw farm assurance schemes as a valuable way of communicating value to consumers. Surveys of public opinion in Europe have highlighted animal welfare as a major public concern and one that should form an essential element of farm assurance. The need to incorporate proper assurance regarding animal welfare formed the basis of the major multinational study titled 'Integration of animal welfare in the food quality chain: from public concern to improved welfare and transparent quality' (Welfare Quality Network, 2009). Both of these initiatives were pre-dated by the RSPCA's 'Freedom Food' scheme, which began in 1994 and is known today as 'RSPCA Assured'.

The aim of all quality assurance (QA) schemes is twofold: to provide an independent audit of welfare standards on farms and to provide assurance to consumers that standards are being met. I have kept up to date with the RSPCA Assured scheme. It cannot, of course, guarantee absolute freedom from any abuse on every farm at any time and it has inevitably come under fire from some who wish to discredit the whole approach but lack positive proposals for any workable alternative. Nevertheless, I retain my belief that this scheme is a power for good, not least because the University of Bristol Animal Welfare group, formerly led by me, was directly involved in setting the standards.

A welfare-based QA scheme should be able to provide evidence to demonstrate that standards of husbandry and welfare on participating farms are consistent with the assurances it claims. As far as the animals are concerned, the aim of good husbandry is to promote a state of well-being, defined most simply as 'fit and feeling good'. The protocols developed for the basis of welfare assessment must therefore include measures of the provisions necessary to establish good husbandry (e.g. resources, records and stockmanship) and most QA protocols have given major attention to these things. However, there is now general agreement that these protocols must include direct, animal-based outcome measures of the physical (fit) and emotional (feeling good) elements of welfare based on sound foundations of animal welfare science.

The development of robust monitoring protocols for husbandry and welfare is an essential first element of welfare-based QA. However, the scheme must also provide good evidence of quality control, namely proof that the monitoring procedure leads to effective action, to ensure overall compliance with required standards, to record any breaches of these standards and to remedy any specific areas where the independent assessor has identified needs for improvement. Moreover, any market-led scheme that seeks to add value on the basis of assured standards of animal welfare surpassing the statutory minimum must ensure that this added value is recognized by both producers and consumers, and is properly apportioned through all links in the food chain.

If customers are to pay more, they need to be aware of, and trust, the assurances provided by the scheme. If retailers are to reward their suppliers for their compliance with superior standards, they too need to promote the scheme to get financial reward through increased market share. If farmers are to invest time and resources to improve animal welfare, they need a financial incentive since most of them are doing the best they can with what they can currently afford. The farm animals, the objects of these good intentions, will only benefit if all three responsible parties can be persuaded to act together. Currently, there is some evidence that some welfare-based QA schemes may be failing to achieve their desired impact (Whay *et al.*, 2003a, b).

Possible reasons for this lack of impact include:

- inadequate monitoring procedures
- failure to develop action plans based on information gathered during the monitoring procedures
- lack of financial incentive for farmers to implement financial plans
- lack of consumer demand for 'high-welfare' produce, arising from a lack of awareness, trust or perceived added quality of individual QA schemes.

If it is to succeed, a welfare-based QA scheme(or the animal-based element of a broader scheme) needs to operate both on the farm and at the retail level. It needs to establish two virtuous cycles: the producer cycle involving the self-assessment, external monitoring, action and review on the farm and the retailer cycle involving the process of QA and quality control. These cycles need to run together as elements of a single, continuous dynamic process. I describe this as a 'Virtuous Bicycle' (Fig. 5.1). I shall describe this whimsical, but conceptually sound, model in detail later. At this stage it is necessary only to indicate that the right wheel of the bicycle illustrates actions relating to the quality of husbandry and welfare on the farm and the left wheel illustrates action to promote the market share for high-welfare products. The direction of the delivery vehicle is towards progressive improvement in standards of animal welfare and progressive increase in customer demand, based on sound evidence to justify the assurance of added value through high welfare standards.

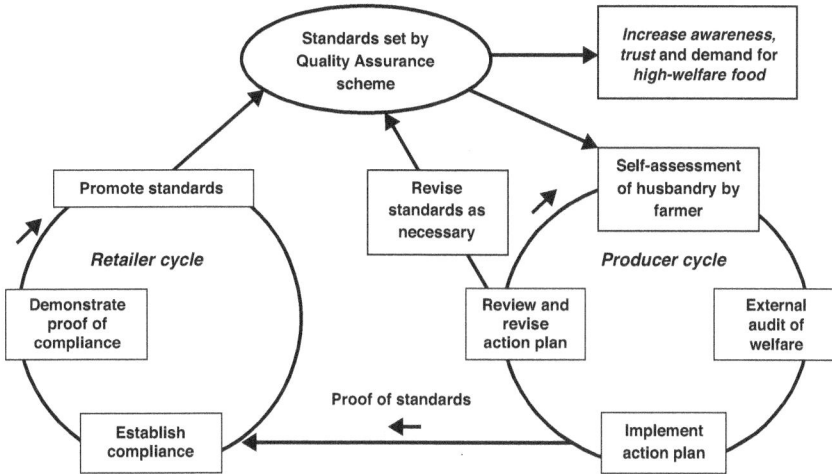

Fig. 5.1. The 'Virtuous Bicycle': a vehicle designed to deliver improved animal welfare on farm. The producer cycle illustrates a dynamic process of self-assessment, external monitoring, action and review on farm. The retailer cycle illustrates the process of quality assurance and quality control at the retailer level. The direction of the bicycle is towards increasing awareness, trust and demand for high-welfare food. (From Webster, 2009, used under licence CC BY.)

The Development of Welfare-based Farm Assurance Schemes

All current European Farm Assurance schemes recognize the need to make proper provision for animal welfare, whatever the primary basis for the insurance. All the protocols acknowledge the Five Freedoms (FFs) as principles by which to define standards of animal welfare. However, while the promotion of these freedoms is a stated aim, most provisions have been based almost entirely on audit of the provision of resources to the animals and records of management procedures, such as the provision of health care. What they lacked was a significant element of outcome measures, i.e. animal-based welfare assessment. It has become generally accepted by those working in welfare science that these observations and records of provisions should be augmented and, in many cases, replaced by animal-based measures. (Bartussek, 1999; Whay *et al.*, 2003a). This has become a central theme for International Workshops on Animal Welfare Assessment at Farm and Group Level (which carries the unfortunate acronym WAFL).

The European Commission, through Framework Programme 6, funded a major, multi-national study titled 'Integration of animal welfare in the food quality chain: from public concern to improved welfare and transparent quality' (Welfare Quality®, 2009). Two of the key aims of this programme were:

- to develop robust on-farm welfare monitoring and information systems for selected farm animal species; and

- to implement a welfare-monitoring and information system and develop welfare improvement strategies.

These aims reflect the fact that QA in the matter of farm animal welfare can only be guaranteed on the basis of robust protocols for the welfare state of the animals, backed up by proof of effective action to implement and sustain husbandry procedures necessary to promote satisfactory welfare and address any specific problem that might arise. The principles that underpin the monitoring protocols have been outlined by Botreau *et al.* (2007a,b).

On-Farm Monitoring Protocols Based on Direct Animal-based Measures

It is necessary to make the distinction between welfare assessment and welfare monitoring.

Welfare assessment

This can involve any or all of the science-based physiological, behavioural or motivational methods used to determine the welfare of a sentient animal as it seeks to cope with challenges to its physical and mental state. Many of these methods have been developed under laboratory conditions and do not readily transfer to on-farm application. Moreover, most have been devised to address specific and ever more subtle elements of welfare state, e.g. the motivation of hens to dustbathe or the motivation of intensively housed pigs to seek fresh air.

Welfare monitoring

In the context of farm animal welfare, this describes the process where trained observers (monitors) seek to build up an accurate impression of overall welfare (or separate categories of overall welfare) in a population of farm animals from a series of observations and measurements according to standard, agreed protocols. These procedures for incorporation into on-farm monitoring protocols must be underpinned by scientifically proven methods for assessment of physical and mental state. They must also be robust, quantifiable and sufficiently objective to minimize between-observer variation. Moreover, any monitoring protocol for a QA scheme will only be acceptable if it can be completed within a day or less and without undue disturbance to the animals or normal farm routines. This inevitably calls for a degree of compromise. Protocols based on animal-based measures taken by an independent observer on a single day also raise the concern that they are no more than snapshots and may fail to reflect the long-term picture. This can be offset to a degree by selecting animal-based measures that integrate long-term consequences of former husbandry.

Table 5.1. The four principles and twelve criteria proposed by Welfare Quality®
as elements of protocols for the direct, animal-based assessment of welfare (from
Botreau *et al.*, 2007a, b, Table used under licence CC BY.)

Welfare principles	Welfare criteria
Good feeding	Absence of prolonged hunger
	Absence of prolonged thirst
Good housing	Comfort around resting
	Thermal comfort
	Ease of movement
Good health	Absence of injuries
	Absence of disease
	Absence of pain induced by management procedures
Appropriate behaviour	Expression of social behaviours
	Expression of other behaviours
	Good human-animal relationship
	Absence of general fear

Development of protocol

Animal-based protocols have been developed for the audit of farm animal
welfare by the team at Bristol University (e.g. in dairy herds, Whay *et al.*,
2003b). Our original aim was to test the former, provision-based protocols
that formed the basis for QA standards set out by the Soil Association and
those by the RSPCA's 'Freedom Foods' scheme. In the dairy cow protocol,
physical and mental health were assessed within six categories: nutrition,
reproduction, disease, external appearance, environmental injuries and
behaviour. The protocol for free-range laying hens included measures of
attitude (arousal and response to a novel object), activity (feather pecking,
aggression and use of range) and physical welfare (mortality, body condition
and egg quality). Selection of the more subjective measurements for inclu-
sion in the protocol was based on consistency of measurement between the
different observers.

The protocols developed by Welfare Quality® are based on similar lines to
those of the Bristol group. This is outlined in Table 5.1. It recognizes four principles of well-being, good feeding, good housing, good health and appropriate
behaviour. These may be assessed on the basis of twelve more specific criteria,
each amenable to direct monitoring under farm conditions. For example, good
housing is defined by the criteria of comfort around resting, thermal comfort
and ease of movement, The specific observations and measurement necessary
to establish these criteria need to be kept under constant review and tested for
accuracy, robustness (absence of significant variation between observers) and
practicality.

Welfare Quality:

Progressive evaluation structure

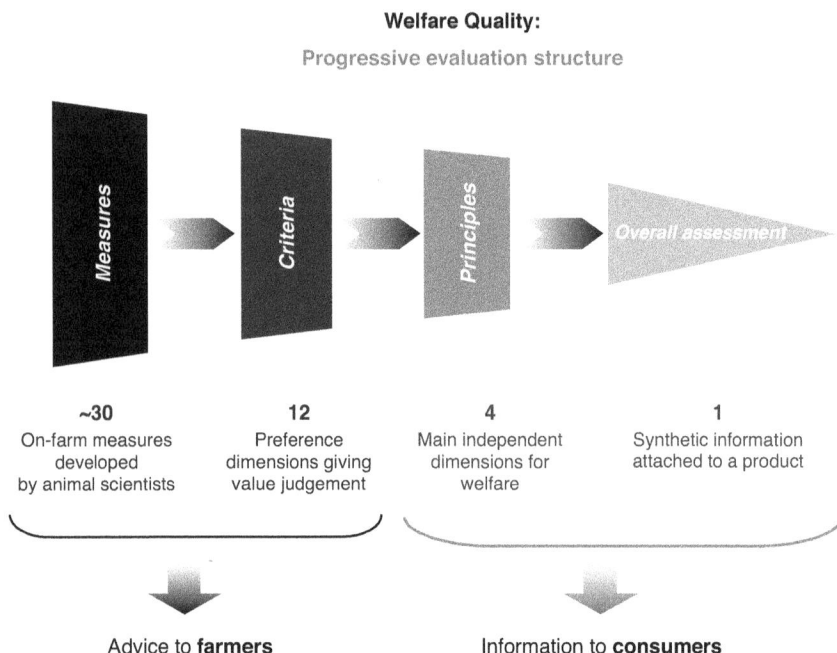

~30	12	4	1
On-farm measures developed by animal scientists	Preference dimensions giving value judgement	Main independent dimensions for welfare	Synthetic information attached to a product

Advice to **farmers**   Information to **consumers**

Fig. 5.2. The Welfare Quality® approach to the characterisation of measures and criteria used for the on-farm monitoring of animal welfare and possible routes for the integration of these measures for conveying information to farmers and consumers. (From Botreau *et al.*, 2007a, b, used under licence CC BY.)

Having developed a protocol based on robust measurements of welfare criteria, categorized in terms of the four principles of welfare quality or the FFs of the Farm Animal Welfare Council (FAWC) (FAWC, 2005), we need then to agree as to how this information can be best used. The degree to which the separate welfare criteria can and should be aggregated depends on how the information is to be used and by whom. If, for example, the aim is to improve welfare on an individual dairy farm by reducing lameness, then the monitor's report should include specific information as to risks arising from hazardous flooring. If, at the other extreme, the aim is to conclude whether a particular unit does or does not meet the standards laid down by a particular QA scheme, then the individual criteria have to be aggregated to produce an overall score or, at least, a category of pass or fail. Such aggregation is necessarily subjective since it relies on value judgements as to the relative importance of the different principles and criteria. If the decision as to compliance or non-compliance is to be made simply on the basis of the monitor's report, then aggregation becomes essential. This is, of course, a very good reason for *not* making conclusions as to compliance simply on the basis of this report.

Welfare Quality® has proposed the following approach to the characterisation, aggregation and interpretation of the animal-base measures incorporated in their protocols (Fig. 5.2). The four principles and twelve criteria of good welfare (Table 5.1) that form the structure of the protocol are used as the basis for value judgements. They are derived from a larger number (perhaps 30) of specific, proven, robust measures of physical and mental state (e.g. body condition, prevalence of lameness, incidence of mastitis, evidence of feather pecking). The twelve criteria, along with the larger number of specific records, can form the basis for a strategic welfare plan for each farm that includes specific decisions for action in relation to feeding, housing and management. Communication to consumers will be based on the four principles and the single overall assessment.

In the Welfare Quality® scheme, it is proposed that farms will be ranked on a four-point scale: unclassified, basal, good and excellent. This approach shows promise, but unresolved issues remain. Questions that arise include 'how good is excellent?' Would it require evidence of positive welfare, (e.g. happiness)? Perhaps the most serious objection to this proposal to rank farms as unclassified, basal, good and excellent will prove unacceptable in a commercial world that operates according to a different language of hyperbole where the word 'best' is more usually interpreted as the minimal standard for acceptance. If there is to be a ranking system, my preference would be for one that was strictly numerical (e.g. 0–3 stars).

Design of the Virtuous Bicycle

However well designed and scientifically proven the assessment protocol, it cannot be expected to succeed unless it leads to effective action on farms to promote and, wherever possible, improve standards of animal welfare. My proposal is that the protocols should be formally incorporated into a continuous process of self-assessment, external monitoring, action and review. This becomes the right wheel of the virtuous bicycle (Fig. 5.1).

Each cycle begins with a structured written assessment prepared by the farmer with input from stock workers and veterinarians as appropriate. The self-assessment will be based on the standards of husbandry and provisions as set out by the QA scheme and will include records of feed provision, housing and hygiene, health, use of medicines, stockmanship and training, the existence and operation of a health and welfare plan. The farmer should also outline any known welfare problems and priorities for action to address these concerns.

The next stage of the cycle is the visit by the independent monitor, trained for and operating according to the standards of the QA scheme. The visit will include an interview with the farmer to assess and discuss and review the self-assessment and an inspection of the animals and provisions for the animals to assess welfare. The report of the monitor will contain an overall assessment of compliance with the standard and strategic advice in relation to specific issues. It should identify areas where it is desirable or necessary to improve welfare and recommend an action plan to address these specific issues. An action plan for the control of lameness in dairy cattle has been described by Bell *et al.* (2009). The farmer will then read the report, address any points of misunderstanding or disagreement and submit a written response before it is submitted to the supervisors of the QA scheme. After an appropriate interval (normally 1 year) the inspection cycle, based on self-inspection followed by independent monitoring, will be repeated.

By starting the producer cycle with self-assessment, farmers can address elements of husbandry, provision and records in their own time, thus reducing the amount of work that has to be done at the time of the visit from the external assessor. It recognizes that farmers have the most knowledge (if not necessarily the best) of the procedures in operation on their own farms and why they have evolved. The aim of the visit by the external assessor is to mount a fair challenge to the self-assessment. While the first visit will be comprehensive, subsequent assessments can concentrate on the most important issues arising from previous assessments and the success or otherwise of the action plan.

The practical merits of this approach are as follows:

- It incorporates elements of husbandry, including the provision of resources and records of actions to promote welfare.
- Once the first cycle of self-assessment, monitoring, action and review has been established, subsequent revolutions of the cycle can focus on the most important issues and avoid bureaucratic and time-wasting repetition of all elements of the protocol.
- Compliance would not normally be based on the results of a single monitoring exercise but on the effectiveness of actions to promote welfare and address specific problems. This reduces the risk of subjective bias in the assessors' reports and variation in standards between assessors. It is also more challenging to farmers because it does not allow them simply to file away the assessors reposts and forget about them until next year. They must provide evidence of effective action.

The left wheel of the virtuous bicycle (Fig. 5.1) represents the retailer cycle and is designed to improve public awareness of, and demand for, food and other products from farms operating to proven high welfare standards. The aim is to create a sustainable, verifiable process of information transfer to retailers and consumers (i.e. all of us) as to welfare standards and actions to ensure welfare standards on farms operating within the QA scheme. These standards should be freely available for inspection by all, both in outline and in detail. Farmers can enter the scheme only when they can provide evidence of action to establish compliance with the standards of the scheme.

Riding the Bicycle: The Delivery Process

The virtuous bicycle cannot be powered by virtue alone. The scheme will inevitably incur greater costs to both producers and retailers than QA schemes based simply on an annual visit of 1 day or less. It is therefore unrealistic to expect it to succeed unless it brings real rewards to all stakeholders, consumers, retailers, farmers and (of course) the animals themselves through proper recognition of added value accruing from proper attention to animal welfare. Produce bearing the logo of this value-added scheme should retail at a higher price than that produced through schemes doing no more than meeting nationally approved (minimum) standards and a fair proportion of this increased price should go the farmer. If it became pan-European policy to impose the monitoring standards and rating system proposed by welfare quality, then it would be logical to equate 'unclassified' (or zero star) as the minimum. Awards of one, two or three stars would generate rewards as increments of added value in terms of animal welfare with commensurate increases in the price of the produce.

The model of the virtuous bicycle has been offered simply as a concept. It is, however, one that emerges from an awareness that current welfare-based QA schemes have a long way to go if we are to achieve significant improvements in animal welfare at farm level and a significant increase in consumer demand for proven high-welfare food. At this stage, I should point out that the scheme is not designed to compel all consumers to buy food at a price determined by those of us who can afford the luxury of compassion. Nevertheless, it does offer a way forward. However, it will only work if both farmers and consumers have trust, based on transparency and evidence that the scheme does what is says on the tin. In short, the virtuous bicycle will only deliver when both wheels turn together.

Welfare-based QA schemes are now a fact of life and are likely to become more and more internationally widespread. The tide of public opinion is in our favour but if we fail to deliver through piecemeal approaches that fail to sustain the revolution of both wheels of the cycle of challenge, action, promotion and reward, then consumers, retailers and farmers could lose faith and set back the cause of animal welfare for years.

Note

[1] This letter has been drawn from the article *The Virtuous Bicycle: A Delivery Vehicle for Improved Farm Animal Welfare* (Webster, 2009).

References

Bartussek, H. (1999) A review of the Animal Index System (ANI) for assessment of animals' well-being in housing systems for Austrian proprietary products and legislation. *Livestock Production Science* 61, 179–192.

Bell, N.J., Bell, M.J., Knowles, T.G., Whay, H.R., Main, D.J. *et al.* (2009) The development, implementation and testing of a lameness-control programme based on HCCP principles and designed for heifers on dairy farms. *The Veterinary Journal* 180, 178–188.

Botreau, R., Veissier, I., Butterworth, A., Bracke, M.B.M. and Keeling, L.J. (2007a) Definition of criteria for overall assessment of animal welfare. *Animal Welfare* 16(2), 225–228.

Botreau, R., Bonde, M., Butterworth, A., Perny, P., Bracke, M.B.M. *et al.* (2007b) Aggregation of measures to produce an overall assessment of animal welfare. *Animal* 1(8), 1179–1187.

Curry Commission (2002) *Report of the Curry Commission on the Future of Farming and Food.* Available at: https://dera.ioe.ac.uk/id/eprint/10178/

Farm Animal Welfare Council (FAWC) (2005) *Report on the Animal Welfare implications of Farm Assurance Schemes.* FAWC, London, UK.

Webster, J. (2009) The virtuous bicycle: A delivery vehicle for improved farm animal welfare. *Animal Welfare* 18, 141–147.

Welfare Quality Network (2009) *Assessment Protocols for Cattle, Pigs and Poultry.* Welfare Quality Consortium, Lelystad, The Netherlands. Available at: https://www.welfarequalitynetwork.net/en-us/reports/assessment-protocols/

Whay, H.R., Green, L.E. and Webster, A.J.F. (2003a) Animal-based measures for the assessment of welfare state of dairy cattle, pigs and laying hens: Consensus of expert opinion. *Animal Welfare* 12, 205–217.

Whay, H.R., Main, D.C.J., Green, L.E. and Webster, A.J.F. (2003b) Assessment of the welfare of dairy cattle using animal-based measurements: Direct observations and investigation of farm records. *Veterinary Record* 153, 197–202.

Compassionate Sustainability: Green Milk from Contented Cows

6

The essential needs of humans and all animals for the right amount of the right sort of food are immediate, continuous and have long-term consequences for the quality of our lives and that of the planet. Those with too little to eat are unable to promote normal development and good health for themselves and their offspring. Those who eat too much accelerate their rate of decay. The Food and Agriculture Organisation (FAO) report (FAO, 2020) states that 9% of the world population (820 million people) are experiencing severe hunger, two billion (22%) experience moderate to severe food insecurity. More than 20% of children under five show stunted growth. At the same time 13% are described as moderately to severely obese. Furthermore, things are getting worse at both ends of the scale.

Increasing numbers of humans self-evidently increases demand for food and puts increasing strain on the capacity of the living environment to meet this demand. However, population numbers, per se, present less of a strain on resources than increased demand from those with money to spend on attractive but environmentally spendthrift sources of food, especially that from animals given food that we could have eaten ourselves. Based on records of population numbers and calorie consumption per capita, I estimate that if the Chinese consumed the same amounts of the same sort of food as eaten by citizens of the USA (as is their right) they would exhaust the resources of the planet in less than 30 years. Radical change in the way we produce, distribute and consume food, especially food from animals, is not just a moral aspiration, it is an ecological necessity.

It is self-evident that food from animals puts a greater demand on resources than food from plants, since so much of that food is required to maintain the animals themselves. This simple conclusion has laid the foundations for a number of complex analyses of resource use such as that featured in the report *Livestock's Long Shadow* (FAO, 2006). Such analyses describe the relative inefficiencies of using animals to exploit resources of sun, soil and water to produce food for human consumption, the degradation of these resources through overuse, the pollution of land and water from excessive waste products and the threat to the climate from greenhouse gases (GHGs), especially methane (CH_4) from ruminants, which has approximately 20 times the global warming potential of carbon dioxide (CO_2).

Public criticism of the scale and practice of current methods of farming animals for food is based on the following four premises:

* most of those who can, consume too much meat and milk
* food that we could eat is fed to animals while the poor go hungry
* livestock's long shadow is destroying the planet
* intensive livestock production is incompatible with animal welfare.

The dairy industry has become a major target for criticism on all four counts. Given the current state of the industry, much of this criticism is justified. However, it was not ever thus and there are sound reasons why it need not be so in the future. As a child at school, I got free milk, not always a treat on a summer's day after it had sat for some hours in the sun. This policy was a consequence of the classic work of Sir John Boyd Orr in the 1920s who demonstrated that much of the differences among children in both growth *and educational attainment* could be resolved by improved provision of essential nutrients provided by milk. For most of us today milk has become just a commodity, cheaper to buy than some brands of bottled water. The expansion of the dairy industry has been driven largely by the increase in demand for luxury foods, butter, cream and a huge variety of ice creams, yoghurts and cheeses. These luxuries are acceptable to lactovegetarians because they do not directly involve the killing of animals.

All this is very recent history. For most of recorded history, getting enough to eat was a struggle for survival and the cow was a highly valued partner in this struggle. The traditional role of the family cow was to provide milk, work, fertilizer, fuel, clothing and the occasional fatted calf for special occasions, while being sustained by fibrous feeds that the family could not digest for themselves, usually from land that the family did not own. She was not competing with the family for food, she was an essential contributor to the harvest. The modern dairy cow, typified by the Holstein breed, is a very different creature: bred, fed and managed to produce as much milk as possible within intensive, highly mechanised dairy units. Meat production has become a relatively minor consideration, with male calves destined for beef or veal sent, more often than not, off farm to other specialist rearing units. Other traditional roles for the family cow have disappeared altogether. The modern Holstein is most unlikely to be harnessed to a plough! However, while the high yielding cow, confined in a barn, has become the norm for much of the urbanized, affluent human population, the proportion of cows kept in such intensive conditions in the less developed nations is relatively low.

Table 6.1 provides an illustration of the range of cow numbers and milk yields in different regions of the worlds. The highest yields are recorded in USA, Israel and Saudi Arabia, In the latter two nations, especially, demand is high but the land and climate are entirely alien to the concept of milk from pasture.

Table 6.1. Total milk production (million tonnes), cow numbers (millions) and average milk yields of cows (litres per cow per year). (Table author's own. Used under Licence CC-BY.)

Region	Total milk production (million tonnes)	Cow numbers (millions)	Average milk yields (litres/cow/year)
USA	87.2	9.1	9600
European Union	135.5	23.0	5900
India	50.3	43.6	1150
New Zealand	17.0	4.6	3700
World	599.4	264.4	2270

The EU embraces a range of production systems, from the highly intensive systems used in (e.g.) Denmark and the Netherlands to pastoral systems used in Ireland and much of Eastern Europe. In the EU, average annual milk yield is 5900 l (range: Denmark 8400 l, Bulgaria 1850 l). New Zealand is included in Table 6.1 as an example of an advanced industry producing milk primarily from grass. Average annual milk yield from the world population of 264 million dairy cows is only 2270 l.

The first message to be drawn from Table 6.1 is that the dairy industry, unlike the broiler chicken industry for example, cannot be viewed, criticized or applauded as a single homogenous system. The dairy cow is a creature of infinite variety that can adapt to a wide range of husbandry methods. This creates the potential to adopt a sustainable approach to all four of the concerns outlined above. The charges on which the dairy industry stands accused: unhealthy, unfair, unsustainable and unkind, are not without substance. However, they are (obviously) simplistic because they do not begin to address all the variables in a complex industry. I shall not attempt to rebut these accusations but will address them objectively and explore possible pathways to greener, kinder solutions.

An Unfair Dairy Industry? Food That We Could Eat Is Fed to Animals While the Poor Go Hungry

This is more than an expression of concern. It is a fact but it is not an inevitability. When considered in nutritional and ecological terms, the costs and benefits of food from animals are influenced by the extent to which they may or may not compete with us for resources. The greatest of the essential demands of animals on resources is for energy from food to fuel the processes of life. This is best described in terms of metabolizable energy (ME) measure in megajoules per kilogram (MJ/kg) of dry matter. This defines the amount of fuel that can be extracted from the diet by the processes of digestion and metabolism. In an adult animal in energy balance, neither gaining nor losing weight, all ME is used for maintenance and converted into heat. ME consumed in excess of maintenance is retained in the body as protein and fat or, in a lactating animal, secreted as milk. Food production from animals is inevitably less efficient than that from plants because much of the food eaten by animals is required to meet their own needs. At maintenance, the gross efficiency of conversion of animal feed into food for human consumption is zero. The net efficiency of conversion of increments of ME fed above maintenance to energy in milk may range from approximately 0.6 to 0.8. The overall *gross* efficiency of conversion of energy in animal feed to energy in animal product (meat, milk or eggs) increases with increasing ME intake to a limit set by physiological constraints on appetite. The limit to appetite in highly productive farm animals grown for meat is about three times maintenance and gross efficiency of conversion about 0.3. High yielding dairy cows may consume ME at five times maintenance and achieve a gross efficiency of 0.5. For further explanation see Webster (2016).

Table 6.2 compares the efficiency of conversion of feed energy (ME) and protein into hens' eggs, cows' milk, pork meat from the offspring of sows giving birth to 22 piglets/year and beef from extensively reared cow-calf systems where the contribution of the breeding beef cow is one calf/year plus her own carcass at eventual slaughter.

In each column of Table 6.2 the efficiency of conversion of feed energy and protein is expressed in two ways.

- Overall efficiency: food energy and protein (for human consumption) relative to total ME and protein consumed by the productive and support animals. For meat animals these correspond to the slaughter and breeding generations. For the dairy industry they correspond to the lactating adults and replacement heifers.
- Competitive efficiency: food energy and protein (for human consumption) relative to animal consumption of ME and protein from 'competitive' feed sources (i.e. feeds such as cereals that could have been fed directly to humans) as distinct from 'complementary' feeds (grazing, forages and by-products remaining after preparation of food and drink for human consumption (e.g. maize gluten, brewers' grains).

The *overall efficiencies* of ME conversion into eggs, pork, milk and beef are 0.33, 0.19, 0.42 and 0.08 respectively, while for protein conversion they are 0.32, 0.25, 0.28 and 0.09. The reason why the efficiency of energy conversion to milk is greater than that for egg production can be attributed to the fact that there has been, to date, no limit to the ability of breeders to select cows to produce more and more milk, whereas hens are still restricted to the production of one e.gg per day. Both milk and egg production are more efficient than the intensive production of pork meat, while beef production fails (by these measures) to achieve an efficiency of 10%.

Table 6.2. Efficiency of energy and protein conversion in meat, milk and egg production (Webster, 2021). For each system, efficiency is described by the ratio of output to input, where output is defined by energy and protein in food for humans; inputs are described in terms of total and 'competitive' intake of ME and protein, where 'competitive' describes energy and protein from feed sources that could be fed directly to humans. (Table author's own. Used under Licence CC-BY.)

	Eggs	Pork	Milk	Beef
Production unit	1 hen	22 pigs	1 cow	1 calf
Support unit	0.05 hens	2 sow	0.33 heifers	1 cow
Output/year (kg food)	15	1300	8000	200
Food energy (MJ)	130	13,000	28,000	2500
Protein (kg)	1.65	208	264	32
Input/year (MJ of ME), total	389	67,038	67,089	29,850
'competitive'	351	53,630	20,127	10,268
Input/year (kg protein), total	5.2	818	946	361
'competitive'	5.0	736	236	108
Efficiency				
Food energy/total feed ME	0.33	0.19	0.42	0.08
Food energy/ 'competitive' feed ME	0.35	0.24	1.39	0.24
Food protein/total feed protein	0.32	0.25	0.28	0.09
Food protein/ 'competitive' feed protein	0.33	0.28	1.12	0.30

When energy conversion is examined in terms of *competitive efficiency* the picture changes. Here beef becomes as efficient as pork (or no less inefficient) and dairy farming becomes very efficient indeed. In this example, based on a typical diet fed to cows in the pasture-rich south-west of England approximately 65% ME is complementary and the output of food energy for human consumption is 39% greater than their demand for feed that we could eat ourselves. The ability of the dairy cow to produce more food for human consumption than she eats is most marked in advanced pastoral systems, such as those seen in New Zealand, but can be achieved in fully housed systems though proper selection of complementary feeds.

Table 6.2 provides a powerful illustration of the fact that it is possible, within modern, highly productive production systems, to exploit the ages-old capacity of the milch cow to contribute, rather than compete, in the constant endeavour to provide good food for ourselves, both rich and poor. It would be a mistake, however, to assume that because it can be done, it is being done. The present state of dairy production, especially in the rich, urbanised nations, involves far too much land to grow crops like energy-rich cereals, and protein-rich beans and seeds to drive dairy cows to produce more milk than is compatible with health and welfare for them, us and the planet at large.

An Unsustainable Dairy Industry? Livestock's Long Shadow Is Destroying the Planet

This is a fiendishly complex issue to address because, by definition, it has to embrace all of life. Attempts to achieve a comprehensive assessment of the inputs, outputs and environmental impact of any biological or industrial process are conventionally based on the principles of life cycle analysis (LCA). The International Organization for Standardization (ISO 14040:2006) defines LCA as 'the study of environmental aspects and potential impacts throughout a product's life cycle from raw material acquisition through production, use and dismissal. Environmental impacts needing consideration include resource use, human health and environmental consequences'. This is easier said than done because, as defined, it includes everything, including much that we cannot measure with any certainty and much that is subjective. Any manageable approach to LCA will posit specific questions and select data that would seem to be most relevant to these questions. When researching the literature on resource use and environmental impact of livestock production systems, it is unsurprising therefore to discover a wide range of conclusions and opinions among authors all using valid scientific methods. This can usually be attributed to the fact that they have posed broadly similar questions but in slightly different ways. My (similarly non-comprehensive) approach to LCA in livestock production systems will focus on two of the most important issues, energy use and carbon (C) balance, especially the net production of CO_2 and other GHGs especially CH_4.

The FAO report *Livestock's Long Shadow* (FAO, 2006) catalogues in great detail the ways and the extent to which livestock production, carried out in the manner and at the scale that exists today, is creating an unsustainable burden on the living environment. Ruminants are singled out for special criticism because of their contribution to global warming through the emission of GHGs, especially CH_4, produced from fermentation of fibrous feeds in the anaerobic environment of the rumen. CH_4 has approximately 20 times the global warming potential of CO_2. The report also considers ways to mitigate this, along with other environmental threats from land degradation from overgrazing, and the pollution of land and water from nitrogenous wastes within current intensive systems operating at current levels of production. A grossly oversimplified take-home message from their conclusions would be that it is best for the environment to eat eggs, poultry and pork reared intensively indoors. In my opinion however, the report falls short on several counts. It dodges the central issue, namely that our current problems arise not from livestock production per se, which has been an integral part of sustainable agriculture for millennia, but that it is the current scale of livestock production, both intensive and extensive, that grossly disrupts the ecological balance. To give an obvious example, nitrogen pollution from agricultural waste is simply a case of too much fertilizer in the wrong place. (One could make the same argument for CO_2 production: since plants need CO_2 for photosynthesis, plants need animals to produce CO_2). The report calculates the environmental cost of production systems in terms of global hectares of land required to produce a standard amount of different foods for humans of plant and animal origin but does not adequately take into account the differing capacity of different classes of land to produce crops, e.g. grasses vs cereals. It does not properly account for such things as differences in the availability and therefore the value of site-specific resources, most especially, water. For example, problems associated with water supply and disposal are very different for dairy units located in Israel and those located in the west of Scotland. It does not fully consider the extent to which the effects of the emission of GHGs may be offset by C sequestration in pasture and woodlands grazed and browsed by ruminants.

My main objection to the FAO report is that while it considers strategies for mitigating environmental costs within the context of current production methods and consumption levels, it gives little attention to the extent to which livestock husbandry, using appropriate species in sustainable numbers can, at best, make a positive contribution to environmental quality or, at least, greatly mitigate the costs.

Life Cycle Analysis: Energy and Carbon Inputs, Outputs and Emissions

Tara Garnett and her colleagues provided an excellent overview of the impact of ruminant production systems on the climate and nitrogen pollution of waterways in the report titled *Grazed and Confused* (Garnett *et al.*, 2017). Their report also gives proper attention to ways in which pastoral systems, properly managed, can enhance the quality of the land.

In this section I apply the principles of partial LCA in an attempt to quantify and compare C and energy exchanges in livestock production systems. The main C and energy inputs are feed and fuel, the main products are food (milk and meat) and 'wastes', principally nitrogenous wastes in manure and GHGs that are released into the atmosphere.

A substantial weight of literature has accumulated in respect to net GHG emissions from livestock production systems. Net GHG emissions describe the algebraic sum of GHG production as CO_2 and CH_4 (mostly from animals and manure) set against C sequestration in land grazed by the cattle. Plants convert atmospheric CO_2 into organic matter by the process of photosynthesis. C is stored in the plant, above and below the ground, so long as the organic matter continues to exist, alive or dead. Selectively felling forest trees to build houses or battleships stores C and gives other trees space to grow and store more. Slashing and burning the jungle to clear space for soya or palm oil production brings double jeopardy: it releases all the C into the atmosphere in the form of CO_2 and radically reduces the future capacity for C storage. Untouched tropical rain forests, where nearly all C is retained within the system as organic matter, sequester C long-term. A high proportion of C captured by photosynthesis is stored as organic matter within the soil. It follows that soil erosion is a major contributor to GHG production. The soil under permanent pastures of mixed grasses and clovers will store much more soil organic C (SOC) than arable land used for intensive production of cereals and oilseeds. However, there is a limit to the amount of C that can be stored so that, in time, an equilibrium is reached where net C exchange between plants and atmosphere is zero.

Most of the carbon-based fuels upon which we depend today were laid down during the carboniferous era. At the beginning of this period atmospheric CO_2 concentration was about 20 times the concentration of the 300 ppm (parts per million) recorded at the beginning of the Anthropocene in about 1850, when human mining and consumption of fossil fuels began its long ascent. The climate at the start of the carboniferous period was hot, wet and most of the planet was under water. Most of the land consisted of tropical rain forest, which sequestered nearly all the C it captured. By the end of the carboniferous period, 60 million years later, atmospheric CO_2 had fallen below 200 ppm. Atmospheric oxygen (currently 21%) was over 30% at that time. Physics dictates that this period of climate instability must have ended in a catastrophe and it did, bringing an ice age.

Table 6.3 presents a condensed and greatly simplified summary of data gathered by Pelletier *et al.* (2010a, b) to illustrate the application of LCA to calculate energy use and production of GHG in meat production systems in the USA. The examples include three intensive (commercial) systems, broiler chickens, pork and feedlot beef and two more 'natural' systems; 'niche' pork (equivalent to organic) and beef cattle finished at pasture. The numbers are expressed to only two significant figures, given the high dependence on assumptions, even this approximation almost certainly implies a greater degree of accuracy than is warranted in terms of the absolute numbers. However, the same rules and assumption are applied throughout so the comparisons between systems may

Table 6.3. Life cycle analysis of energy inputs and emissions of greenhouse gases (measured as CO_2 equivalents) in the production of 1 kg of meat in broiler chicken, pork and beef production systems (Pelletier *et al.*, 2010a, b). (Table author's own. Used under Licence CC-BY.)

Output (1 kg meat)	Energy use (MJ) Feed Fuel Total	GHG (kg CO_2 equivalents)
Broiler chicken	10 5.0 15	1.3
Pork, commercial	6.1 4.9 11	2.7
Pork, niche	7.1 5.9 13	3.2
Beef, feedlot finished	28 10 38	35
Beef, pasture finished	41 7 48	46

be treated with some confidence. While the less intensive systems may have relied to a greater extent on complementary feed sources, fuel energy costs were significantly greater. GHG production was conspicuously greater from beef cattle finished at pasture than in feedlots.

These comparisons should be treated with caution since they are specific to the production methods that they describe and cannot be applied world-wide. The high fuel costs for pasture-finished beef cattle in the USA reflect the high dependence on nitrogenous fertilizers. Nevertheless, they illustrate the important point that more 'natural' methods are likely to be less sustainable *according to these criteria* mainly because slower-growing animals have a lower gross efficiency of utilisation of ME (as described above) and produce more GHG equivalents per tonne of meat for human consumption. However, this analysis, like all partial LCAs, is based on limited, selected premises. It does not, for example, take into account the impact of organic farming methods on soil quality, ecological diversity including sentient wildlife or, of course, the welfare of the farmed animals.

There is a weighty volume of literature on GHG emissions from dairy cattle. CH_4 production is a consequence of anaerobic fermentation in the rumen, thus most of the effects of productivity and nutrition can be derived from first principles. Increasing individual milk yield decreases the amount of GHG emissions produced per litre milk as the proportion of digestible energy directed to milk production increases with respect to that required for maintenance (see Table 6.2, this Letter). Higher yielding dairy cows are fed a diet containing a higher proportion of starch to cellulose in the diet and this increases fermentation to propionate relative to acetate. This reduces the proportion of fermentable energy that is lost to the system in the form of 'excess' protons converted to CH_4. These basic principles are explained in greater depth in *Understanding the Dairy Cow* (Webster, 2020).

The most effective ways to maximize milk production relative to GHG production are to feed high energy rations to high genetic potential cows. Because this is, of course, the way to maximize income it is already common practice, which means that there may be little scope for further gains. The use of feed additives such as seaweed or pharmaceuticals to manipulate the rumen microbiome and reduce numbers of methanogens (e.g. archaea) is a conceptually sound approach and a high research priority. This is too big a subject to review here. Suffice it to say that some approaches may have promise, but it is too early to say how well initial inhibition of CH_4 production can be sustained. There is some evidence that it may be possible to reduce GHG production through genetic selection of cows, but even if this is proven to be correct, progress is likely to be slow and slight.

GHG emissions associated with milk production may be compared with those associated with an equivalent production of food from simple-stomached animals. Rotz *et al.* (2010) used data from dairy herds in California and Pennsylvania with annual yields ranging from 5500 to 11,000 kg/lactation to calculate values for GHG production of 0.4–0.7 kg CO_2 equivalents per litre of milk produced. To compare these values with those in Table 6.3, the energy value of 1 kg of meat may be taken as approximately three times that of 1l of milk, so that GHG emissions from milk production correspond to 1.5–1.8 kg CO_2 equivalents per kg meat equivalent. By this measure, the GHG impact of milk production is intermediate between that of chicken and pork. Beef production, by any means, is extremely profligate.

Few, if any, of the soils in land currently used for agriculture are likely to be in a state of C equilibrium. Evidence based on measurements of soil organic C show that while much of the arable land used for the intensive production of cereals and oilseeds is losing carbon, European grasslands (for example) are currently sequestering C, thus acting as a sustained C sink (Soussana *et al.*, 2010). These estimates of net C balance (CO_2 equivalents per m^2 land) in European pastoral systems for dairy and beef production predict that the rate of C sequestration relative to GHG production increases with the proportion of feed that is directly grazed, so that by this measure extensive beef production from cattle fed at pasture becomes the most sustainable.

Emergy Analysis

Because all forms of life cycle analysis are partial, they will inevitably lead to different conclusions according to the questions asked and the variables included in the model. To my knowledge, the closest approach to a comprehensive LCA of exchanges of energy and matter in any production system is that known as 'emergy analysis', where emergy (Em) is a measure of the amount of the original, effectively inexhaustible source of solar energy embedded at each stage of the process This concept expresses all the work processes and resources (sunlight, water, fossil fuels, minerals etc.) used in the generation of a product in terms of a common unit of measurement (Zhao and Li, 2005). The approach is fiendishly complex and, like most LCAs, carries a lot of uncertain assumptions but it is, I believe, particularly well-suited to the assessment of the efficiency and sustainability of farming the land for food because it can identify, distinguish and quantify the renewable (R) resources of sun, soil and water embedded in farmland from non-renewable (NR) resources and purchased resources such as fuel, fertilizer, labour and imported feeds (F).

In the context of food production, resources are defined as follows:

- renewable Emergy (R) = emergy equivalents from sustainable resources, e.g. sunlight, free water;
- unrenewable Emergy (UR) = loss of emergy from (e.g.) soil degradation;
- purchased goods and services (F) = bought in feed, fuel, labour, etc.; and
- yield (Y) food for human consumption.

Table 6.4 compares yields and sustainability in different agricultural systems on the basis of the following ratios.

Table 6.4. Yield and sustainability within agricultural systems assessed in terms of embedded energy ('emergy') and described by three ratios, emergy yield ratio (EYR), environmental load ratio (ELR) and emergy sustainable index (ESI) These ratios are dimensionless. For further explanation see text. (Table author's own. Used under Licence CC-BY.)

	EYR	ELR	ESI
Maize (USA)[a]	1.07	18.8	0.06
Conventional pig (Sweden)[a]	1.04	22.3	0.05
Organic pig (Sweden)[a]	1.13	7.80	0.15
Intensive dairy (Brittany)[c]	1.35	3.25	0.42
Extensive dairy (Mali)[c]	1.89	1.25	1.57
Grazing cattle (Argentina)[b]	3.73	0.55	6.80

[a]Pereira and Ortega, 2013
[b]Rotolo *et al.*, 2007
[c]Vigne *et al.* 2013

- EYR (emergy yield ratio) = (R + NR + F)/F. This describes the contribution of local resources (land) to product.
- ELR (environmental load ratio) = (NR + F)/F. This describes the ratio of non-renewables to renewables in product.
- ESI (emergy sustainable index) = EYR/ELR. This becomes a measure of yield relative to environmental compatibility)

Values for EYR show that the relative contribution of local renewable resources did not differ greatly between maize production, conventional and organic pig production. The contribution of local resources was greater for dairy production, especially low-intensity dairy production in South Mali. This is consistent with the evidence presented previously in Table 6.2. The grazing of beef cattle was by far the most efficient in respect to the contribution of renewable resources. The most striking differences between the systems are revealed in the ESI column, the measure of yield in relation to environmental compatibility. By this measure, small scale dairy production is more sustainable than intensive production, even in Brittany where a large proportion of feed comes from pasture, and extensive beef production on the Argentinian pampas outstrips all others in terms of sustainability. This may come as a surprise to urbanized critics of livestock production and beef production in general, but production and beef production in general, but it would appear as an overcomplicated proof of the obvious to the gauchos of the pampas or the indigenous races of North America living in perfect symbiosis with the bison.

The approach to calculating the environmental costs of agricultural systems included in *Livestock's Long Shadow* was based primarily on land use and concluded that the environmental cost of feeding people on beef is 10 times the cost of cereals and 40 times the cost of soya. The emergy approach yields a diametrically opposite conclusion (Table 6.4). By this analysis and in this example, maize and soya are the least sustainable because of their dependence on non-renewable resources of (e.g.) fertilizer and fuel (F) and degradation of soils (UR); beef from cattle grazing the pampas of Argentina are the most sustainable, both in terms of food emergy yield relative to the consumption of non-renewable resources (F/NR) and in terms of overall sustainability, defined by the ESI.

I concede that the examples illustrated in Table 6.2 have been chosen by me to make a point. Different approaches tell different stories. However, they all point to the same two conclusions. The first is that the current demand for foods of animal origin, particularly when this involves the feeding to animals of food that we could have eaten ourselves, is unsustainable. The second conclusion is that the key to sustainable farming is to manage different land types in ways that best respect the value of the location and land as defined by its own special resources of sun, soil and water. This, indeed, is the essence of husbandry. Nobody, I hope, would consider ploughing up the Argentine pampas. Nobody, at least, who is aware of the disastrous consequences of ploughing up the North American prairies that led to the dustbowl of the 'dirty thirties'.

It makes good ecological sense to derive value from land best suited to pastures through the production of food of high nutritional value from animals dependent, so far as possible, on complementary feeds that we cannot eat ourselves. It makes even more ecological sense in silvopastoral systems where food production is just one of several contributors to value; others being income from sustainable forestry, water management, habitat and wildlife conservation and, not least, greater C sequestration. These forms of good husbandry cannot, however, produce meat and milk in the quantities that the comfortable and affluent have come to expect.

An Unkind Dairy Industry? Intensive Dairy Production Is Incompatible with Animal Welfare

In recent years, the most common expressions of public concern as to methods of food production have related to issues of farm animal welfare. A particular target for criticism has been the highly intensive systems identified by Harrison (1964) and Brambell (1965). Egg production in the UK is now based largely on 'free range' systems to the satisfaction of the general public, if not necessarily the birds. Today, one of the most serious expressions of public concern is that at a time when we have freed hens from the battery cage and given them free range, the dairy industry has taken the cows out of the fields and confined them on concrete. This is an oversimplified image, but it is a powerful one that needs to be addressed.

I have, in previous letters, described how the publication of the Brambell report (Brambell, 1965) led to the formation of the Farm Animal Welfare Council (FAWC) and the publication of the Five Freedoms and Provisions. These recommendations have stood the test of time. They are measures of *outcome*, now recognized as the most direct approach to the assessment of animal welfare. They are, moreover, not intended as a counsel of perfection but as a guide to good husbandry, being simple enough to be memorable but comprehensive enough to be effective. It is necessary, however, to make the distinction between animal welfare and wellbeing. Welfare describes the physical and mental state of an animal across the whole spectrum from very good to very bad. Wellbeing describes a state within the range of satisfactory to good and must therefore be the aim of good husbandry. When measured strictly in terms of (short-term) economics, large industrial dairy units have been an undoubted success. When measured in terms of the wellbeing of the animals and the land, achieved through sympathetic and sustainable husbandry, they are found to be wanting. The needs that drive the mind of the modern, highly bred, intensively fed cow are much the same as for any sentient mammal: food and water, comfort, security and a stable social life consistent with the genetic imperative for reproduction. Fundamental to all these specific needs is freedom of choice – to take action to avert discomfort or threat and promote a positive sense of wellbeing. As we know too well, the impact of food on our state of mind is not just a matter of acquiring sufficient nutrients. So too with cows. The acts of eating and, in their case, ruminating, bring their own satisfaction. Grazing animals in the wild state have adapted to seasonal changes in food availability: lots of good grass in the summer or rainy season, much less food of much poorer quality in the winter or dry season. It is entirely natural for grazing animals to lose weight during the lean months, but provided some grazing is available, however poor the quality, they get the satisfaction of freedom to forage for what they can.

The most severe welfare problems for the dairy cow are likely to be associated with physical stresses to her physical fitness rather than denial of behavioural expression. Relative to most farm animals she is most unlikely to suffer in consequence of having nothing to do all day. On the contrary she is worked quite extraordinarily hard. The modern dairy cow can cope in the short term with the intense metabolic demands involved in the production of 60l milk/day (or more), coupled with the demands of consuming and digesting enough food to meet these demands. It is an inescapable fact, however, that too many succumb too soon to the long-term stresses of lactation, in particular, the *production diseases* such as rumen acidosis, ketosis, environmental mastitis and lameness that are, by definition, linked to the methods employed in the breeding, feeding and housing of cows to produce large quantities of milk and therefore, by definition, our fault.

Table 6.5 presents a brief summary of potential welfare abuses that may occur in dairy systems. For the most part it is based on the template laid down by the five freedoms, but includes a further stress, namely that of *exhaustion* arising from failure to cope, in the long term, with the exacting physical demands of lactation. For the dairy cow, exhaustion is probably the biggest problem of all. It describes a cow broken down in body, and probably in spirit, through a combination of stresses associated with nutrition, housing, hygiene and management exacerbated in many cases by breeding programmes that

Table 6.5. Abuses of the five freedoms that can arise through systematic failures in the provision of good husbandry. (Table author's own. Used under Licence CC-BY.)

Hunger	Nutrition fails to meet the metabolic demands of lactation
Chronic discomfort	Ruminal indigestion Poorly designed cubicles, inadequate bedding
Pain and injury	Claw disorders (sole ulcer, white line disease) Digital dermatitis Damaged knees and hocks
Disease	Mastitis, ketosis,
Fear and stress	Rough handling, bullying, separation from calf
Lack of choice	Zero grazing, inadequate rest time
Exhaustion	Emaciation, infertility

have overemphasized productivity at the expense of robust good health. Too many infertile, emaciated or chronically lame cows are culled prematurely because they are no longer making a productive contribution to the enterprise. This is not only an abuse of welfare but also a terrible waste since a dairy cow needs to complete at least four lactations to recoup the cost of rearing her as a heifer until she delivers her first calf and enters the milking herd.

The most common breed of dairy cow in intensive systems is the Holstein. During the period 2002–2014, average lactation yields in UK Holsteins increased by 21% from 7637 to 9239 kg/head. Within 'elite' dairy herds in the USA average lactation yields in excess of 11,000 kg are commonplace. These increases have been achieved through a combination of selection strategies heavily weighted towards increased production of milk solids and developments in nutrition designed to support the high metabolic demands of lactation within the constraints of appetite. In simple terms, this involves increasing the ratio of cereals, where the main energy source is starch, and protein-rich oilseeds (e.g. soya, rapeseed) to forages (fresh and conserved grasses) where the main energy source is digestible fibre. Whatever their genetic potential, it is only possible to achieve these high yields if the cows are confined and forage intake is restricted. This policy inevitably presents threats to health and welfare.

Cows are not motivated to eat by a desire to reward the farmer with as much milk as possible, but by the desire to attain a feeling of comfortable satiety. Their capacity to take in food, especially fibrous food essential for healthy digestion, is constrained by the rate at which this food can be fermented in the rumen. Selection for increased yield increases the probability that they will be unable to meet their metabolic demand for nutrients to sustain lactation and body condition within the limit of appetite set by the capacity and rate of digestion within the rumen. If they cannot eat enough to meet their metabolic demands, they will experience a sense of chronic metabolic hunger. To increase nutrient intake within the constraints of gut fill it is necessary to increase fermentation rate within the rumen by increasing the proportion of rapidly digestible starch to slowly digestible fibre. This increases the risk of ruminal acidosis, which is, at least, uncomfortable and, in severe cases, can lead to severe malaise and even death. Many high yielding cows can simultaneously suffer from chronic hunger and the discomfort of ruminal indigestion. This is not a good feeling.

Cows' need for comfort is greatly influenced by their size and shape. The modern Holstein weighs over 700 kg and has prominent joints, especially knees and hocks. For comfort they need to lie down on pasture or a deformable bed of straw or sand. Concrete does not feel good. Cows are motivated to lie down to rest for about 11 hours per day. There comes a point where the need to lie down overrides the need to eat. In many intensive units high yielding dairy cows are milked three times daily, having queued to enter the milking parlour. They are also compelled to eat for at least 8 hours to meet their nutrient demands. With so much to do, the time to lie at rest will be much less than they would wish.

Cows, like all sentient animals, are motivated by curiosity and caution. Curiosity is a powerful motivator in early life as calves seek to gather useful information. In later life, in a stable environment, caution becomes the wiser approach to ensuring a sense of security. Most cows in stable groups establish a stable hierarchy, through the exchange of social signals that usually avoid physical conflict. In houses where each cow has access to an individual cubicle, it is normal for each to use the same cubicle every time. Overworked by the demands of lactation, they opt for the quiet life. However, they do retain their curiosity. If you wish to be entirely surrounded by curious cows, lie down in a field and the rest will follow. Horizontal, we present no threat and become interesting.

Whether on the family farm or in large intensive units, the dairy cow is a valuable individual and will be given individual attention. Despite this, dairy cows are at high risk of three major health problems, infertility, mastitis and lameness. These conditions are known as production diseases, a phrase that concedes that they are largely our fault. Pryce *et al.* (1997) explored the genotypic and phenotypic links between selection for increase milk yield and the incidence of these three conditions. At that time there was a significant genotypic correlation between milk yield and all three. In the case of infertility and mastitis, there was no significant phenotypic link, which indicates that farmers were able to offset genotypic deterioration in these traits through improvements in management. In the case of lameness both genetic and phenotypic correlations were significant, which suggests that farmers were failing to hold the line. In recognition of the genetic link between selection policies being heavily weighted towards increased yield, breeding companies have reformulated their selection indices to give increased emphasis to traits defined as robust as measured by an increase in productive lifespan. In the selection index currently used by the UK Independent Dairy Breeding Company, nearly 70% of traits are now based on measures of fitness, longevity and good welfare. The impact of selection for this set of traits on the progeny of tested bulls is integrated in the form of the *Profitable Lifetime Index*. However, individual farmers can select bulls to suit individual cows and their individual systems by giving individual attention to specific traits related to resistance to the main production diseases: fertility, body condition, locomotion and somatic cell counts (SCC), for resistance to mastitis.

There is, at present, no evidence to suggest that the incidence of production diseases is greater in large intensive units than on the traditional family farm. The incidence of infertility is linked to poor body condition, itself a consequence imbalance between the metabolic needs of lactation and the capacity of the cow to ingest and digest feed. While digestive disorders, especially rumen acidosis, are a major threat to the welfare of dairy cows, improvements in understanding of ruminant nutrition and the application of this new knowledge to the formulation of total mixed rations have done much to reduce the risks attached to the selection and management of high genetic merit cows to produce prodigious quantities of milk. The risk of physical discomfort, pain and injury in dairy cows attributable to poor housing and inadequate control of lameness is high. However, once again, there is no convincing evidence to suggest that these problems are worse in large, intensive units where cows are confined throughout lactation than in small family farms, where cows are at pasture during the summer. Indeed, the physical environment within large, new, expensive dairy units can often present a lower risk of injury than on the traditional, old, undercapitalized family farm.

There are some practices that we inflict on cows entirely for our benefit, in full knowledge that they conflict with how they would naturally perform to promote a sense of wellbeing. The top three, in ascending order of importance, are:

- tethering cows throughout the time they are housed;
- keeping cows permanently housed, without access to pasture; and
- removing calves from their mothers shortly after birth.

In many small rural communities it has been traditional to keep dairy cows outdoors all summer on lush pastures, like Alpine meadows, then bring them in for the winter and tether them in tie-barns where they will be fed, watered and milked until turn-out in the spring. This practice has given rise to concern mainly on the grounds that it denies freedom of movement and opportunities for a social life. I know of no evidence that cows display signs of distress associated with prolonged tethering, although passing the winter group-housed in a barn with deep clean straw and access to an outside yard would undoubtedly be better. Some free-stall houses with insufficient, poorly bedded cubicles and filthy passageways can be worse than tie-stalls. In any event, tie-stalls are incompatible with modern milking systems and will, I predict, gradually fade away.

In a few areas, such as UK, Ireland and New Zealand people are accustomed to seeing dairy cows outdoors at grass during the summer, so assume this to be the natural state. However, this is becoming the exception through most of the developed world where the majority of lactating dairy cows are kept off pasture throughout their working life. The trend in commercial dairy production, worldwide, is towards industrialized units of 1000 or more very high-yielding cows. In order to sustain these high yields, the cows are housed throughout lactation and given continuous access to a ration that ensures they take in far more nutrients than they could possibly derive from grazing at pasture because nutrient density of the ration is greater and the feed can be consumed more rapidly. Confinement also keeps the cows close enough to the milking parlour to permit thrice-daily milking or, increasingly, the use of a robot milking machine that they can enter of their own free will. This offers freedom of choice. However, robot milking machines are only practicable when cows are permanently housed. Mature cows do not appear to be strongly motivated to enter the milking parlour simply to relieve discomfort to their distended udders. They need a food stimulus to attract them in from pasture. The attractions of pasture can be greater than the attractions of the robot milker, even when feed is on offer in the parlour. In consequence they visit the robot less often and milk yield falls. This makes it progressively easier for the cow to meet her metabolic needs from pasture so increases her preference to stay outdoors. Her welfare will improve but her productivity will fall. In some large, intensive units, cows are confined throughout lactation but given a period of recreation on pasture for a few weeks during the dry period when they have completed their lactation and await their next calf.

Pasture provides an excellent source of nutrients in the form of fresh and conserved grasses and clovers. Moreover, when the weather is fine, pasture is an ideal environment for dairy cows. Here they can do much as they please: take in food, excrete urine and faeces, exercise, rest, enjoy fresh air and space, socialise and satisfy their curiosity. There is however a conflict between the use of pasture as a recreation area and the need to maximize its potential as a source of high-quality feed. Once the first flush of spring grass is over, most of the best grasslands are harvested for silage. For much of the so-called grazing season cows may be turned out onto 'sacrifice' pastures that provide little nutrition but all the other amenities. In these circumstances pasture is serving only as a recreation area. The most cow-friendly farm I have ever seen was in the forested foothills of the Pyrenees in northern Spain. Cows could choose to roam in comfort and security among the trees, or rest in well bedded kennels. There was little of nutritional value in the forest but much of interest. Nutrition, including freshly cut grass in season was provided at a feeding station close to the milking parlour. This 'zero-grazing' system came as close as possible to meeting all their day-to-day needs, but it was exceptional. Ideally, cows should have freedom of choice to go outdoors, when they wish, where there is space, cool fresh air and a comfortable place to rest. Cows are, undoubtedly highly motivated to graze fresh grass and I am always moved to watch the scenes of excitement when they are first turned out in spring, but I cannot find strong evidence to indicate that they suffer from the inability to graze, per se.

Our most extreme disturbance to the emotional state and natural behaviour of dairy cows is the policy of removing their calves shortly after birth, partly for ease of management but mainly to maximize income from sale of milk. It is difficult to estimate the possible magnitude of this practice in scientific terms. We have little option but to consider the practice within the context of the natural behaviour of cow and calf.

Whether in the wild, or out-of-doors on the farm, the natural behaviour of the dairy cow at parturition is to separate from the herd and give birth in what she thinks will be a safe spot, for example, close to a hedge. Having licked the calf into shape and given it a first meal, she leaves it and returns to the herd. She instructs the calf, in effect, to lie still and unnoticed until she returns to give it another drink. This behaviour is hard wired and has survival value. After a few days, when the calf has become active and can move as well as its mother, it will join the herd, spending much of the time with other calves, because they are more interesting, visiting its mother perhaps four to six times daily for a feed and usually resting with her at night. It is natural for cows and their calves to spend a long time apart, but both show signs of distress if not together at mealtimes. A few farmers separate cow and calf but allow the calf to join its mother twice daily to take a modest feed before the rest of the milk goes into the machine. This system appears to be acceptable to both mother and calf. Many domesticated water buffalo, e.g. in India, will not permit themselves to be milked unless their calf is present.

While I believe that the twice-daily access system is a reasonable approach to sympathetic husbandry, it is likely to remain a minority pursuit. What then is the least-worst approach to early weaning? In this context, the French word *sevrage* is more accurate. At present, the most common practice is to separate the calf within 24 hours of birth. On some traditional dairy farms, calves will be left with their mothers for two to 3 weeks to ensure they get the full benefits of mother's milk. However, weaning after 3 weeks undoubtedly causes more distress to both cow and calf than weaning shortly after birth. Early weaning is an unpleasant business but, in the words of the murderous Macbeth, 'when 'tis done, it were well it were done quickly'.

The Poison Is in the Dose

I have, so far, sought to address major sources of criticism of modern dairy practice arising from both the concerned general public and those with professional knowledge of the industry. In each case, I start from the premise that there is a case to answer then proceed to examination of the evidence. Some is taken from new science, and here I have included a small number of citations, as an introduction to further reading. However, most of my argument has been based on established scientific principles of nutrition, physiology, genetics and behaviour, together with equally well-established practical principles of good husbandry. Selected references to original scientific communications would, I believe, add little to this element of my argument.

The common theme that emerges from examination of this critique of the dairy industry is that problems are almost entirely problems of scale: 'the poison is in the dose'. Most of us who can, consume too much meat and milk for our own health and for the health of the planet. A significant reduction in our consumption of food from animals, especially those that are largely dependent on food that we could eat ourselves (e.g. cereals and proteinaceous beans and seeds), would greatly reduce the amount of land needed for growing crops and thereby improve the long-term quality and sustainability of the land through reforestation, rewilding and C sequestration, especially within the soil. A diet and lifestyle that excludes all food and other products of animal origin may be ethically justified within a framework that considers ethics only within the human dimension but becomes difficult to justify when considered within the broader context of efficient use of resource and sustainable management of the ecosystem, especially the huge areas of natural grasslands and savannah (grasses, trees and shrubs). At present, much of this land has been degraded by overgrazing. However, well-managed pastoral and silvo-pastoral systems can improve the quality of the land as measured in terms of plant and animal diversity, soil quality, C sequestration and amenity value. Conservation grazing, using a stable population of suitably adapted ruminants involving a sensible programme of population control, can be an essential to this approach to sustainable land management, sustained, in part, through a policy of controlled culling of animals for human consumption. This can be more profitable and more humane than leaving them out to starve or be eaten by wolves.

Pollution of the soil and water with agricultural wastes from intensive livestock units is, I repeat, a case of too much potentially valuable fertilizer in the wrong place. In the case of pollution with nitrogenous materials, much of this arises from a non-renewable resource bought into the unit in the form of fertilizers and high-protein feed supplements, and disposed of at too high a concentration, too close to the factory farm. The core principle of organic farming is to ensure the maximum possible contribution, recycling and conservation of resources derived from within the farm itself. In the short term, this can never generate yields to compare with production units that depend wholly or in large part on purchased, non-renewable resources. In the long term, however, they offer the only truly sustainable option.

Currently, CH_4 production from ruminants is estimated to contribute approximately 10% to the planetary production of GHGs. The current cattle population of the USA is (very approximately) 100 million animals, of which about 40 million are adult cows. It has been estimated that in the 17th century, before the arrival of Europeans bent on slaughter, the bison population of North America was approximately 60 million. After adjusting for the fact that grazing animals produce more CH_4 per unit of digestible energy than cattle fed on high concentrate rations, one can make a rough estimate that CH_4 production from ruminants in North America is only about 15% higher than it was 300 years ago. If, as seems inevitable, we are compelled to reduce world production and consumption of meat and milk by 20–25%, then levels of CH_4 production from ruminants should return to pre-industrial levels.

Moreover, as explained earlier, most of the accusations that cows are wrecking the planet fail to consider the extent to which the effects of CH_4 production may be mitigated by C sequestration, especially in situations where ruminants derive their sustenance entirely (or almost entirely) from permanent pastures. Well-managed grasslands can constitute a significant C sink, the extent of C sequestration depending on factors such as the intensity of grazing and the balance between grasses and legumes. The true impact of ruminants on climate change through the net production of GHGs can only be determined by LCA of the production and sequestration of GHGs (CO_2 and CH_4) in different systems. While it is the case that estimates based on LCA show that all current dairy systems make a positive net contribution to GHG production, it is far less than estimates based on CH_4 emissions alone and least of all when the contribution of pasture to the overall diet is greatest. Extensive systems of beef production from pasture are likely to be GHG neutral (as would have been the herds of prairie bison).

Opportunities for Change

We cannot escape the fact that our present rate of consumption of foods of animal origin is unsustainable. It is in our own interests to embark now on a strategic programme of change in livestock farming with similar aims to our current long-term programme to work towards net C balance. Indeed, the two strategies overlap within the same overriding, essential need, i.e. to restore the balance of nature. Unless we make some relatively painless changes to our lifestyle now, our children will have far more uncomfortable changes thrust upon them in the future. However, we will not (in sufficient numbers) do this of our own free will while the status quo remains so comfortable. We must be made to change. This will require a balanced menu of attractive carrots and humane applications of the stick.

This is a big subject. My brief is restricted to changes that can be achieved within the dairy industry. Any strategy for change must take account of, and give proper respect to, the needs of the consumers, the farmers, the environment and the cows. It must also plan for an absolute reduction in global milk production. This is counter to current economic thinking that continuous growth is essential to economic stability. In biological terms, this premise is, of course, an absurdity and was well expressed by David Attenborough who said 'the only people who believe in continuous growth are economists and lunatics'.

The prospect of new, greener, kinder approaches to milk production becomes more realistic when we reflect that the hyper-intensive dairy industry in the affluent industrialized regions of today's world is not the norm, but a product of the last 50 years, an intense but unsustainable spike in the balance of nature. I reprise my words at the outset, for most of recorded history the role of the family cow was to provide milk, work, fertilizer, fuel, clothing and the occasional fatted calf for special occasions, while being sustained by fibrous feeds that the family could not digest for themselves, usually from land that the family did not own. She was not competing with the family for food; she was an essential contributor to the harvest and she was valued accordingly.

I am not suggesting that we should return to 'the good old days', not least because for most people dependent on subsistence agriculture then and now throughout most of the underdeveloped world, days were and are not that good. What I am saying is that any future developments should incorporate all that is of value in new knowledge and technology but also ascribe proper value to the sources of this wealth, the cows and the land. Respect for cows may be a moral issue, respect for the land is a matter of survival. These principles apply equally throughout the dairy industry from the highly intensive >1000 cow dairy units of Wisconsin to the dairy syndicates of India receiving and process-ing milk from multiple small famers, each with perhaps only two to five cows.

Increased sustainability of food production systems depends on increas-ing the contribution of renewable as distinct from non-renewable resources. I have briefly described an elegant and comprehensive way to quantify these by way of 'emergy' analysis. This makes it possible to estimate (with considerable uncertainty) an ESI for different systems.

Table 6.4 turns current agricultural economics on its head. Maize, which ranks highest in terms of productivity (yield/ha) becomes the worst when measured in terms of sustainability. Beef cattle, sustained entirely from pasture are the least productive, but most sustainable. This is an extreme illustration of a general truth, which is that increased sustainability of food production from animals must be accompanied by a reduction in production. This has to be a good thing for the health and welfare of ourselves (the consumers), the cows and the living environment. It can however present serious problems for farmers and consumers, particularly those with least money to spend,

I have the good fortune to live in Somerset, classic cow country from time immemorial. The word 'Somerset' describes the land of the summer people, who brought their cattle down each spring to graze the coastal marshes, which flooded in winter but were a reliable source of quality pasture throughout the driest of summers. Table 6.2, which shows that dairy cows can produce 40% more food energy for human consumption that they consume in terms of food that we could eat ourselves, is based on data taken from the feeding programme of my immediate neighbour, who grows over 60% of the feed for his cows on farm. The largest producers of yoghurt in the UK farm the Somerset grasslands to organic standards. A central tenet of their policy has not been to select their cows for milk production per se, but for milk production from pasture, which inevitably means less milk per cow.

The dairy industry in New Zealand is almost entirely pasture-based but presents cows (and their calves) many of the stresses associated with the most intensive indoor systems. Cows are expected to calve at 12-month intervals to synchronize peak lactation and peak grass supply. In 1960, 60% of the dairy herd were Jerseys. Thereafter, genetic selection almost entirely favoured Holsteins based on criteria similar to those applied to Holsteins in the USA bred to live in barns. A selection index heavily weighted for milk yield was, in this environment, incompatible with maintaining high fertility at 12-month intervals. For some time, there was a policy to abort cows that were slow to conceive. Thankfully, this policy has largely been abandoned, reflecting a selection policy designed to place greater emphasis on fertility.

While the production of milk from grazed pasture can be an excellent example of good husbandry, farming the land for what it is best equipped to provide and selecting the cows best suited to this policy, it represents a small and diminishing sector of the international dairy industry. The greater challenge is to apply the principles of sustainability to the vast numbers of cows kept on large, industrialized units with little, if any, access to pasture. In theory, it would be possible to provide a high proportion of feed from local renewable resources (e.g. organic grassland). In practice, short-term economics dictate that most producers will rely to a large extent on bought-in feed and fertilizer. This leads to problems of waste disposal, especially nitrogen and phosphorus. The European Union has issued directives to limit emissions of nitrogen and phosphorus, reinforced by levies for exceeding defined limits. Dutch dairy farmers have responded to these directives by reducing the application of nitrogen and phosphorus fertilizers to grasslands and, in some cases reducing protein in concentrate rations. This application of the stick has reduced pollution problems at the 'cost' of a small reduction in productivity as measured by lactation yield.

There are, at least in theory, several approaches to the reduction of CH_4 emissions from rumen fermentation. As explained earlier, CH_4 emission relative to milk production falls with increased milk yield and increased intake of starchy concentrates. Moreover, when cows are housed and fed on a formulated total mixed ration there is greater potential to reduce CH_4 emissions through control of diet and manipulation of the ruminal microbiome. It is also possible to reduce total emissions of GHGs and other pollutants such as nitrates through improved manure management. Llonch *et al.* (2017) have reviewed the health and welfare consequences of alternative approaches. The bulk of the evidence suggests that it may be possible to reduce CH_4 production by up to 50% through a combination of diet and the use of drugs such as ionophores. However, any manipulation of the rumen population designed to depart from the 'normal' carries the risk of reducing fermentation rate and thereby feed intake.

A particularly attractive solution to problems of CH_4 emissions (and much else) is the development of Silvo-pastoral systems where cattle graze, browse and relax within a parkland area of pasture, shrubs and trees that act as shelters and C sinks (Cubbage *et al.*, 2012; Chará *et al.*, 2019). In Brazil, for example, there are highly successful commercial silvo-pastoral systems (both beef and dairy), that generate income both from the cattle and the sale of tree biomass. A different but equally attractive example of ecologically sound diversification can be seen in the cork-oak parklands of Portugal grazed by the beautiful Mertolenga cattle. Income is generated from the sale of beef, corks for high-quality wines and tourists wishing to enjoy the natural environment. The cattle can select what to graze or browse and where to lie to their satisfaction (e.g. in sun or shade). In all but the most severe weather, they are comfortable and, above all, have freedom of choice.

These examples show that there are ways to produce green milk and meat from contented cows, but they are the exception. The important question addresses how we may aspire to these aims within the great majority of industrialized high input/high output systems. Short-term economics that measure success simply in terms of profit margins will always favour the most intensive system. Some control over this can be achieved through imposition of penalties for environmental pollution, but greater progress can be achieved through a judicious selection of carrots.

Happily, in recent years, public pressure for higher animal welfare standards and political pressure to mitigate environmental costs have started to move things in the right direction. One approach is to farm to organic standards set by the Soil Association (2025) that require (e.g.) no use of artificial fertilizers and that a minimum of 60% of the ration should be based on fresh or conserved pasture. At present only 4% of dairy farms in the UK are organic. However, these farms are competing successfully because there is a niche market for organic milk. As I write, the average price for organic milk is about 40 p/litre; conventional milk about 30 p/litre. By contrast, oat milk, with a much lower nutritional value retails at about £1.40/litre.

Public demand for high standards of cow welfare has had a greater impact than the demand for organic milk, probably because the financial cost to consumers has been relatively small. Thanks largely to public pressure for higher welfare standards, most dairy herds in UK now operate according to the standards set by a Welfare Quality Assurance Scheme. Examples include the Red Tractor Scheme, RSPCA Assured (formerly Freedom Foods) and those established by competing supermarkets. All require monitoring by independent assessors to ensure compliance with the standards of the scheme. This is not the place to argue in detail about the relative merits of the different schemes. However, those operated by the supermarkets have had the largest uptake in terms of milk sold. This is an example of how competition within the free market can be a force for change. Supermarkets recognize a public demand for higher animal welfare standards, albeit somewhat price-elastic, and compete by including on their shelves, products of animal origin, like milk and free-range eggs produced according to quality-assured high welfare standards. The aim is to attract customers to this supermarket on the basis of these assurances who then do the bulk of their food shopping in the same store. This allows the supermarket to pay a higher price for quality-assured milk, without significant effect on their overall profit margins.

While the incentives and penalties considered above are steps in the right direction, they fall far short of the changes needed to achieve the aim of 'green milk', where 'green' may be defined by net zero GHG emissions. This should be incorporated into the aims of the International Climate Commission and lead to government action enforced by law. It is now generally accepted (if not in general practice) that agricultural subsidies should be redirected from support for food production towards support for public goods such as long-term management of soil and water resources, C sequestration, diversity of habitat and wildlife conservation. This would recognize that farmers are, by default, not only food suppliers but are also the most important direct custodians of the natural environment. This is a lofty aspiration. It remains to be seen how close we shall get to meeting this aim and whether the money involved will be sufficient to achieve significant improvement in environmental quality without bankrupting farmers in the process. In the specific context of green milk from contented cows, it has the potential to address two of the cows' greatest challenges, overwork and lack of choice. The stress of overwork can be reduced through feeding and breeding strategies designed to achieve the more robust cow, producing less milk per lactation, but with a longer, more comfortable, productive life. The problem of lack of choice can be addressed by ensuring that in any policy of environmental enrichment for the public good, the word 'public' should embrace the cows.

Notes

[1] This letter has been drawn largely from the article *Green Milk from Contented Cows: Is it Possible?* (Webster, 2021).
[2] The article contains a comprehensive list of references to original communications. I have restricted the references listed below only to those that I believe you may find necessary to the understanding of this letter.

References

Brambell, F.W.R. (1965) *Report of the Technical Committee of Enquiry into the Welfare of Animals Kept Under Intensive Husbandry Systems.* Cmnd.2836, HMSO, London, UK.

Chará, J., Reyes, E., Peri, P., Otte, J., Arce, E. *et al.* (2019) *Silvopastoral Systems and their Contribution to Improved Resource Use and Sustainable Development Goals: Evidence from Latin America.* FAO, CIPAV and Agri Benchmark, Cali, Colombia.

Cubbage, F., Balmelli, G., Bussoni, A., Noellemeyer, E., Pachas, A.N. *et al.* (2012) Comparing silvopastoral systems in eight regions of the world. *Agroforestry Systems* 86, 303–314.

Food and Agriculture Organisation (FAO) (2020) *The State of Food Security and Nutrition in the World.* Available at: www.fao.org/state-of-food-security-nutrition/en/

Food and Agriculture Organization (FAO) (2006) *Livestock's Long Shadow, Environmental Issues and Options.* Available at: https://www.fao.org/4/a0701e/a0701e.pdf

Garnett, T., Godde, C., Muller, A., Röös, E., Smith, P. *et al.* (2017) *Grazed and Confused.* Food Climate Research Network, Oxford, UK.

Harrison, R. (1964) *Animal Machines: The New Factory Farming Industry*. Vincent Stuart, London, UK.

ISO 14040:2006 (2006) *Environmental Management - Life Cycle Assessment - Principles and Framework*. International Organization for Standardization.

Llonch, P., Haskell, M.J., Dewhurst, R.J. and Turner, S.P. (2017) Current available strategies to integrate greenhouse gas emissions in livestock systems: An animal welfare perspective. *Animal* 11, 274–284.

Pelletier, N., Lammers, P., Stender, D. and Pirog, R. (2010a) Life cycle assessment of high and low profitability commodity and niche production systems in the upper mid-western United States. *Agricultural Systems* 103, 599–608.

Pelletier, N., Pirog, R. and Rasmussen, R. (2010b) Comparative life cycle environmental impact of three beef production strategies in in the upper mid-western United States. *Agricultural Systems* 103, 380–389.

Pereira, L. and Ortega, E. (2013) A modified footprint method: The case study of Brazil. *Ecological Indicators* 16, 113–127.

Pryce, J.E., Veerkamp, R.F., Thompson, R., Hill, W.G. and Simm, G. (1997) Genetic aspects of common health disorders and measures of fertility in Holstein-Friesian cattle. *Animal Science* 65, 353–360.

Rotolo, G.C., Rydberg, T. and Liebline, G. (2007) Emergy evaluation of grazing cattle in Argentina's Pampas Agriculture. *Ecosystems and Environment* 119, 383–395.

Rotz, C.A., Montes, F. and Chianese, D.S. (2010) The carbon footprint of dairy production systems through partial life cycle assessment. *Journal of Dairy Science* 93, 1266–1282.

Soil Association UK (2025) *Organic Standards for Great Britain*. Available at: https://www.soilassociation.org/our-standards

Soussana, J.F., Tallec, T. and Blanfort, V. (2010) Mitigating the greenhouse gas balance of ruminant production systems through carbon sequestration in grasslands. *Animal* 4, 334–340.

Vigne, M., Peyraud, J.L., Leconte, P. and Corson, M.S. (2013) Emergy evaluation of contrasting dairy systems at multiple levels. *Journal of Environmental Management* 129, 44–53.

Webster, J. (2016) *Animal Husbandry Regained: The Place of Farm Animals in Sustainable Agriculture*. Earthscan, Routledge.

Webster, J. (2020) *Understanding the Dairy Cow*. Wiley, Oxford, UK.

Webster, J. (2021) Green milk from contented cows: Is it possible? *Frontiers in Animal Science* 2021(2). DOI: 10.3389/fanim.2021.667196.

Zhao, S. and Li, W. (2005) A modified method of ecological footprint calculation and its application. *Ecological Modelling* 185, 65–75.

Postcard: A Pig's Eye View of Human Behaviour

If we are to understand other animal species, we should try to imagine how we appear to them. To this end, I offer an extreme exercise in reverse anthropomorphism: to imagine how a pig must feel it is to be a human. Wittgenstein (1953) suggested that if a tiger could talk, we wouldn't understand it, presumably because of profound differences in the nature of our consciousness. However, I see no harm in trying. I choose a pig because, as Churchill said, 'Dogs look up to us. Cats look down on us. Pigs treat us as equals'.

I, the pig, start from the premise that humans have as much right as me to be treated as a sentient animal. Sentient animals have feelings that matter. I, the pig, assume that because humans, like us, are sentient you are powerfully motivated, like us, to behave in ways designed to make you feel good and avoid feeling bad. From what I observe, this can involve a wide range of behaviours, some of which I can understand, seeking food, comfort, sex. Others that I don't understand include running marathons, adopting stray dogs, sitting indoors all day discussing the meaning of words like deontology, when the sun is shining and a wallow is at hand.

Being a pig, I am a libertarian but respect the golden rule 'do as you would be done by' (which, if I thought about it, is presumably what you mean by deontology). I therefore acknowledge that you humans should have the freedom to do what you like so long as your actions do not frighten the horses or compromise the freedoms of others. Because I am a well-read pig, I would describe these as negative freedoms. Positive freedoms are permissible only up to a point. In the neat phrase of George Monbiot, 'your freedom to swing your fist stops at the point of my nose'. So far as I, the pig, am concerned you have an obligation to respect four of the five freedoms you have proposed for us, namely the negative freedoms from hunger and thirst, thermal and physical discomfort, pain, injury and disease, fear and stress. The fifth, positive freedom to exhibit normal behaviour is a positive freedom and therefore negotiable. I share with you the grudging acceptance that I should not be free to copulate whenever I like and with whoever is at hand. Having reached this point, I, the pig, conclude that this is about as far as I can go. I humbly concede that, while you humans share with me the property of sentience, feelings that matter, most human behaviour is so alien to my understanding of the world as to be fundamentally uninteresting or constitute a threat. Thus, on balance, I decide that my contact with humans should be kept to the bare minimum consistent with my personal interest.

Ah, you will say, not all animals think and feel like pigs. Sentience and cognitive ability will be governed by species, evolution, education and environment. We can agree on this. I think we can also agree that wild animals will sensibly conclude that they should stay out of your way and you should stay out of theirs. We observe that the domestic animals that you have made dependent on you appear to interact with you in a positive way. However, this may be simply because they anticipate that you are bringing them something they need. Keith Kendrick (1998) recorded brainwaves from sheep. The image of a human triggered a signal of aversion. The image of a human carrying a sack of corn triggered a signal of attraction. You humans should accept that while we respond positively to you when you are being useful to us, you have no right to expect us to love you. So long as I am in a decent habitat with the company of my own kind, contact with humans is not a behavioural need.

I would ask you to extend this concept to relationships with your favourite pets, dogs, cats and horses. Remember they are not the same: dogs have owners, cats have staff. Cats enjoy your company but on their terms. When did your cat last stroke you? Dogs may appear to love you but that is because many of them are not fully aware that they are dogs and that's a problem. Dogs and horses display a broader repertoire of behavioural disorders than any species of farm animal raised in the company of others of its own species. The most common abuse of the welfare of dogs and horses is that they are denied freedom of natural behaviour achieved by social contact with their own kind. This is direct conflict with the UK Codes of Practice for the Welfare of Farmed Animals. If dogs and horses were not classified as pets, this denial of natural behaviour could be considered an abuse.

To conclude: I concede that you have made yourself necessary to some of us, but you remain a pretty inexplicable species. The bit that I can understand is that you, like us, are driven by your sentience and motivated largely by self-interest. This may cause you to feel love for certain animals, individuals and species. However, this does not give you the right to expect us to love you. It follows that the best interactions between humans and other animals are likely to be working relationships based not on emotion but mutual respect for each other as individuals: cowboy and horse, shepherd and sheepdog. I am not remotely concerned as to how you feel, it is what you do that matters.

You may find all this hard to take, and indeed, profoundly disagree with me; but then, you are not a pig.

I rest my case.

Note

[1] This postcard is taken from pages 212–217 in *Animal Husbandry Regained: The Place of Farm Animals in Sustainable Agriculture* (Webster, 2013).

References

Kendrick, K.M. (1998) Intelligent perception. *Applied Animal Behaviour Science* 57, 213–231.

Webster, J. (2013) *Animal Husbandry Regained: The Place of Farm Animals in Sustainable Agriculture*. Earthscan, Routledge.

Wittgenstein, L. (1953) *Philosophical investigations*. Blackwell, Oxford, UK.

www.ingramcontent.com/pod-product-compliance
Lightning Source LLC
Chambersburg PA
CBHW042315210326
41599CB00038B/7136